CAMBRIDGE L

Books of enduring scholarly value

Travel and Exploration

The history of travel writing dates back to the Bible, Caesar, the Vikings and the Crusaders, and its many themes include war, trade, science and recreation. Explorers from Columbus to Cook charted lands not previously visited by Western travellers, and were followed by merchants, missionaries, and colonists, who wrote accounts of their experiences. The development of steam power in the nineteenth century provided opportunities for increasing numbers of 'ordinary' people to travel further, more economically, and more safely, and resulted in great enthusiasm for travel writing among the reading public. Works included in this series range from first-hand descriptions of previously unrecorded places, to literary accounts of the strange habits of foreigners, to examples of the burgeoning numbers of guidebooks produced to satisfy the needs of a new kind of traveller - the tourist.

The Natural and Moral History of the Indies

The publications of the Hakluyt Society (founded in 1846) made available edited (and sometimes translated) early accounts of exploration. The first series, which ran from 1847 to 1899, consists of 100 books containing published or previously unpublished works by authors from Christopher Columbus to Sir Francis Drake, and covering voyages to the New World, to China and Japan, to Russia and to Africa and India. This volume, first published in 1880, translates the first detailed description of the geography and indigenous culture of South America, written by Joseph de Acosta (1540–1600) and published in Spanish in 1590. Acosta was one of the first explorers to record and analyse the geophysical phenomena of the 'New World' and attempt to explain them scientifically. Volume 2 contains Books 5–7 of Acosta's work, describing the life of the indigenous population and including a brief history of the end of the Inca empire.

Cambridge University Press has long been a pioneer in the reissuing of out-of-print titles from its own backlist, producing digital reprints of books that are still sought after by scholars and students but could not be reprinted economically using traditional technology. The Cambridge Library Collection extends this activity to a wider range of books which are still of importance to researchers and professionals, either for the source material they contain, or as landmarks in the history of their academic discipline.

Drawing from the world-renowned collections in the Cambridge University Library, and guided by the advice of experts in each subject area, Cambridge University Press is using state-of-the-art scanning machines in its own Printing House to capture the content of each book selected for inclusion. The files are processed to give a consistently clear, crisp image, and the books finished to the high quality standard for which the Press is recognised around the world. The latest print-on-demand technology ensures that the books will remain available indefinitely, and that orders for single or multiple copies can quickly be supplied.

The Cambridge Library Collection will bring back to life books of enduring scholarly value (including out-of-copyright works originally issued by other publishers) across a wide range of disciplines in the humanities and social sciences and in science and technology.

The Natural and Moral History of the Indies

Volume 2: The Moral History

Joseph de Acosta
Edited by Clements R. Markham

CAMBRIDGE UNIVERSITY PRESS

Cambridge, New York, Melbourne, Madrid, Cape Town, Singapore,
São Paolo, Delhi, Dubai, Tokyo

Published in the United States of America by Cambridge University Press, New York

www.cambridge.org
Information on this title: www.cambridge.org/9781108011525

© in this compilation Cambridge University Press 2009

This edition first published 1880
This digitally printed version 2009

ISBN 978-1-108-01152-5 Paperback

This book reproduces the text of the original edition. The content and language reflect
the beliefs, practices and terminology of their time, and have not been updated.

Cambridge University Press wishes to make clear that the book, unless originally published
by Cambridge, is not being republished by, in association or collaboration with, or
with the endorsement or approval of, the original publisher or its successors in title.

WORKS ISSUED BY

𝕮𝖍𝖊 𝕳𝖆𝖐𝖑𝖚𝖞𝖙 𝕾𝖔𝖈𝖎𝖊𝖙𝖞.

———◆———

THE NATURAL AND MORAL HISTORY
OF THE INDIES.

VOL. II.

No. LXI.

THE

NATURAL & MORAL

HISTORY OF THE INDIES,

BY

FATHER JOSEPH DE ACOSTA.

REPRINTED FROM THE
ENGLISH TRANSLATED EDITION OF EDWARD GRIMSTON,
1604.

AND EDITED,

With Notes and an Introduction,

BY

CLEMENTS R. MARKHAM, C.B., F.R.S.

VOL. II.

THE MORAL HISTORY

(BOOKS V, VI, AND VII).

LONDON:
PRINTED FOR THE HAKLUYT SOCIETY.

MDCCCLXXX.

COUNCIL

OF

THE HAKLUYT SOCIETY.

COLONEL H. YULE, C.B., PRESIDENT.
ADMIRAL C. R. DRINKWATER BETHUNE, C.B. } VICE-PRESIDENTS.
MAJOR-GENERAL SIR HENRY RAWLINSON, K.C.B.
W. A. TYSSEN AMHERST, ESQ., M.P.
REV. DR. G. P. BADGER, D.C.L.
J. BARROW, ESQ.
WALTER DE GRAY BIRCH, ESQ.
E. H. BUNBURY, ESQ.
THE EARL OF DUCIE.
CAPTAIN HANKEY, R.N.
LIEUT.-GENERAL SIR J. HENRY LEFROY, C.B., K.C.M.G.
R. H. MAJOR, ESQ.
REAR-ADMIRAL MAYNE, C.B.
COLONEL SIR WM. L. MEREWETHER, C.B., K.C.S.I.
DELMAR MORGAN, ESQ.
ADMIRAL SIR ERASMUS OMMANNEY, C.B.
LORD ARTHUR RUSSELL, M.P.
THE LORD STANLEY OF ALDERLEY.
EDWARD THOMAS, ESQ.
LIEUT.-GENERAL SIR HENRY THUILLIER, C.S.I.

CLEMENTS R. MARKHAM, C.B., HONORARY SECRETARY.

CONTENTS OF THE SECOND VOLUME.

		PAGE
ANALYTICAL TABLE OF CONTENTS	. . .	i

THE MORAL HISTORY.	Fifth Book	. . . 298
,, ,,	Sixth Book	. . . 390
,, ,,	Seventh Book	. . 448

| CONTENTS OF THE INDEX | . . . . 535 |
| INDEX . | . . . . . 535 |

NOTICE.

The two Volumes are paged throughout, and the Index is at the end of the Second Volume.

ANALYTICAL TABLE OF CONTENTS.

MORAL HISTORY.
BOOK V.

	PAGE
Prologue to the following Books	296

CHAPTER 1.—*That the pride and malice of the Divell hath been the cause of idolatrie.*

Scriptural evidence of the Devil's pride	298
He hath retired into remote parts	299
Always inventing idolatries	300

CHAPTER 2.—*Of many kinds of idolatry the Indians have used.*

Worship of elements called *Huaca*	301
Idolatry of images or of the dead	301

CHAPTER 3.—*That the Indians have some knowledge of God.*

Peruvians acknowledge a Supreme Being	301
There is no name for God among them	302
Offerings to Viracocha by Peruvians	303
Difficulty in dealing with idolaters	303

CHAPTER 4.—*On the first kinde of idolatrie, upon naturall and universall things.*

Adoration of the sun	303
Worship of thunder and lightning	304
Of the earth, rainbow, and stars	305
Names of stars	305
Mexican worship of Vitzilipuztli	305
The sin of idolatry	306

b

ANALYTICAL TABLE OF CONTENTS.

CHAPTER 5.—*Of the idolatry of the Indians used to particular things.*

The Devil has also made them worship base things - 307
Peruvians worship all things in nature - - 308
Belief in a deity of all best things of their kind - - 308
The *Apachitas* or votive heaps - - 309
An Ynca sceptic touching the sun - - - 310

CHAPTER 6.—*Of another kind of idolatry upon the dead.*

Mourning for the dead becomes idolatry - - 311
Images and mummies of the dead - - - 312

CHAPTER 7.—*Of the superstitions they used to the dead.*

Peruvian belief in a future state - - - 313
They carefully preserved dead bodies - - - 313
Human sacrifices to the dead - - - 314
Food and clothes placed with the dead - - - 315

CHAPTER 8.—*Of the manner of burying the dead among the Mexicaine and sundrie other nations.*

In Mexico the priests interred the dead - - 315
Custom of burning the dead - - - 315
The servants of a great man killed at his funeral - 316

CHAPTER 9.—*The fourth and last kinde of idolatry the Indians used, especially the Mexicaines, to images and idolls.*

The Holy Ghost condemns worship of images - 317
The idol of Vitzilipuztli in Mexico - - - 318
Temple of the Mexican idol - - - - 319
Dress and form of the image - - - 320
Idol of Quetzaalcoatl in Cholula - - 321
Feast in Tlascala - 322

CHAPTER 10.—*Of a strange manner of idolatry practised amongst the Mexicaines.*

Intended victims worshipped as the idol - 323

CHAPTER 11.—*How the Divell hath laboured to make himself equall unto God, and to imitate him in his sacrifices, religion, and sacraments.*

The Devil strives to imitate the religion of God	324
Scarce anything that the Devil has not counterfeited	325

CHAPTER 12.—*Of the temples that were found at the Indies.*

In every province of Peru there was a chief Huaca	325
The great temple at Cuzco	326
The golden image of the sun	326

CHAPTER 13.—*Of the Proud Temples at Mexico.*

Famous temple of Vitzilipuztli	327
Description of the temple	328
Other temples in Mexico	329

CHAPTER 14.—*Of the Priests and their offices.*

Mexicans had several grades of priests	330
Their duties	331

CHAPTER 15.—*Of the Monastery of Virgins.*

Many convents of virgins in Peru	332
Duties of the virgins	332
Religious women in Mexico	333

CHAPTER 16.—*Of the Monasteries of religious men that the Devill hath invented for superstition.*

Letters from Jesuits in Japan touching the Bonzes	334
In Peru there were no monasteries for men	335
In Mexico there were houses of secluded men	336
Little boys as novices in Mexico	336
Dress and penance of Mexican monks	337

CHAPTER 17.—*Of Penance and Strictnes the Indians have used at the Divell's persuasion.*

Penance of Mexican priests	339
Discipline at feast of Tezcatlipuca	339
Peruvian fasts	339

iv ANALYTICAL TABLE OF CONTENTS.

CHAPTER 18.—*Of the sacrifices the Indians made to the Divell, and wherefore.*

Peruvian sacrifices of shells, coca, etc.	340
Sacrifice of animals	341
Sacrifices of first fruits	343

CHAPTER 19.—*Of the sacrifices they made of men.*

Human sacrifices in Peru	344
Malice of the Devil	345

CHAPTER 20.—*Of the horrible sacrifices of men which the Mexicans used.*

Mexicans sacrificed captives	346
The manner of sacrificing	347
Dress of the priests	348
The bodies cast down a flight of steps	349

CHAPTER 21.—*Of another kind of sacrificing of men which the Mexicaines used.*

The flaying of men	350
The victim reverenced as a God	351
Captives sought for to sacrifice	352

CHAPTER 22.—*How the Indians grew weary and could not endure the cruelty of Sathan.*

The Indians desire to be freed from priestly yoke	352
Spaniards resolved to abolish the sacrifices	353
A victim spoke after his heart was cut out	353

CHAPTER 23.—*How the Divell hath laboured to imitate and counterfeite the Sacraments of the Holy Church.*

Solemn feast of Raymi in P-	354
Feast of Situa	355
Ceremony resembling communion	355

ANALYTICAL TABLE OF CONTENTS.　　v

CHAPTER 24.—*In what manner the Divell hath laboured to counterfeite the feast of the Holy Sacrament and Communion used in the Holy Church.*

Mexican virgins make the image of their God of paste and honey - - - - - 356
Procession of the idol - - - - 357
The paste offered to the idol - - - 358
After the ceremony the paste is eaten - - - 359

CHAPTER 25.—*Of Confessors and Confession which the Indians used.*

Confession most general in the Collao - - - 361
The Ynca only confessed to the sun - - - 361
Practices of sorcerers - - - - 362
Confession in Japan - - - - 363

CHAPTER 26.—*Of the abominable unction which the Mexicaine Priestes and other nations used, and of their witchcraftes.*

Mexican priests anointed - - - - 364
Method of preparing the unction - - - 365
Reputed power of the unction - - - 366
Sorcerers in Peru - - - - 367
Their divination - - - - 368

CHAPTER 27.—*Of the ceremonies and customs of the Indians which are like unto ours.*

Their baptisms and marriages - - - 369
Mexican marriage and divorce - - - 370
Numerous idols in Mexico and Peru - - - 371

CHAPTER 28.—*Of some Feasts celebrated by them of Cusco, and how the Divell would imitate the mysterie of the Holy Trinitie.*

Peruvian feast of Raymi - - - - 372
Peruvian Trinity - - - - 372
Other months and feasts of the Peruvians - - 374
Feast of Situa - - - - 375

CHAPTER 29.—*Of the feast of Jubilee which the Mexicaines celebrated.*

Feast of Tezcatlipuca - - - 377 to 384

CHAPTER 30.—*Of the Feast of Marchants which those of Chobetecas celebrated.*

Feast of Quetzalcoatl - - - - 384
Mexican festivals - - - - - 385

CHAPTER 31.—*What profit may be drawne out of this discourse of the Indians superstitions.*

Profit from the study of Indian superstitions - 388

BOOK VI.

CHAPTER 1.—*That they erre in their opinion which hold the Indians to want judgment.*

Ill-treatment of the Indians - - - 390
Authorities for Peru and Mexico - - - 391

CHAPTER 2.—*Of the method of computing time and the Kalendar the Mexicaines used.*

Mexican calendar - - - - 392

CHAPTER 3.—*How the Kings Yncas accounted the yeares and moneths.*

Peruvian calendar - - - - 395

CHAPTER 4.—*That no nation of the Indies hath beene found to have had the use of letters.*

Letters unknown in America - - - 397

CHAPTER 5.—*Of the fashion of letters and bookes the Chinois used.*

Chinese writing 399

CHAPTER 6.—*Of the schools and universities in China.*
Chinese learning - - - - 401

CHAPTER 7.—*Of the fashion of letters and writings which the Mexicaines used.*
Mexican picture writing - - - 403
Mexican records - - - 404

CHAPTER 8.—*Of Registers and the manner of reckoning which the Indians of Peru used.*
Peruvian Quipus - - - - 406

CHAPTER 9.—*Of the order the Indians holde in their writing.*
Various ways of writing - - - - 408

CHAPTER 10.—*How the Indians dispatched their messengers.*
Use of messengers - - - - 409

CHAPTER 11.—*Of the manner of government and of the Kings which the Indians had.*
Government in various countries - - - 410
Mexico and Peru compared - - - 411

CHAPTER 12.—*Of the Government of the Kings Yncas of Peru.*
Ynca ceremonies - - - - 412
Ynca government - - - - 413

CHAPTER 13.—*Of the distribution the Yncas made of their vassals.*
Divisions of the Ynca Empire - - - 414

CHAPTER 14.—*Of the edifices and maner of buildings of the Yncas.*
Ynca edifices - - - - - 415
Ynca bridges - - - 416

CHAPTER 15.—*Of the Yncas revenues, and the order of Tributes they imposed upon the Indians.*
Ynca revenue and tribute - - - - 418
Flocks of llamas - - - - 420

ANALYTICAL TABLE OF CONTENTS.

CHAPTER 16.—*Of arts and offices which the Indians did exercise.*
Arts and handicrafts — 421
Head dresses — 422

CHAPTER 17.—*Of the posts and Chasquis the Indians did use.*
Posts and messengers — — 423

CHAPTER 18.—*Of the justice, lawes, and punishments which the Yncas have established, and of their marriages.*
Ynca marriages — 424

CHAPTER 19.—*Of the Originall of the Yncas, Lords of Peru, with their conquests and victories.*
Indian Governments — — 426
Extent of the Ynca Empire - — 427
Origin of the Yncas 428

CHAPTER 20.—*Of the first Ynca, and his Successors.*
Lineage of the Yncas 429

CHAPTER 21.—*Of Pachacuti Ynca Yupanqui and what happened in his time unto Guaynacapa.*
Ynca traditions - - — 430
Manners of the Yncas - - 432

CHAPTER 22.—*Of the greatest and most famous Ynca called Guanacapa.*
Conquests of Huayna Ccapac 433
The body of Huayna Ccapac sent to Lima — 434
Death of Huascar . 434

CHAPTER 23.—*Of the last Successors Yncas.*
Manco Ynca - — 435
Execution of Tupac Amaru - 435
Succession of Yncas - - 436

CHAPTER 24.—*Of the manner of the Mexicans common-weal.*
Mexican succession - — — 436

ANALYTICAL TABLE OF CONTENTS. ix

CHAPTER 25.—*Of the titles and dignities the Indians used.*
Mexican nobility - - - - 438

CHAPTER 26.—*How the Mexicaines made war, and of their orders of Knighthood.*
Mexican warfare - - - - 440
Mexican knighthood - - - - 441

CHAPTER 27.—*Of the great order and diligence the Mexicaines used to instruct their youth.*
Schools in the Mexican temples - - - 442
Training for soldiers and priests - - - 443

CHAPTER 28.—*Of the Indians feasts and dances.*
Dances in Peru - - - - 444
Dances in Mexico - - - - 445
Music and dancing - - - - 446

BOOK VII.

CHAPTER 1.—*That it is profitable to understand the action of the Indians, especially the Mexicans.*
Profit to be derived from history - - - 448

CHAPTER 2.—*Of the ancient inhabitants of New Spaine, and how the Navatlacas came thither.*
Chichemecas, the first inhabitants of Mexico - - 449
Habits of the wild people - - - - 450
Invasion of the Navatlacas - - - - 451

CHAPTER 3.—*How the six Lineages of Navatlacas peopled the land of Mexico.*
Lineages of the Navatlacas - - - - 452
War between Tlascaltecas and Chichimecas - - 453
Peopling of America - - - - 455

CHAPTER 4.—*Of the Mexicaines departure, of their journey, and peopling of the Province of Mechoacan.*

Migration of the Mexicans - - - 456

CHAPTER 5.—*Of that which happened in Malinalca in Tula, and in Chapultepec.*

Continued migration of the Mexicans - - - 459

CHAPTER 6.—*Of the Warres the Mexicaines had against them of Culhuacan.*

Mexicans and the King of Culhuacan - - - 461
Continued march - - - - 462

CHAPTER 7.—*Of the foundation of Mexico.*

Foundation of Tenoxtitlan or Mexico - - 463
Division of the city into quarters - - - 464

CHAPTER 8.—*Of the sedition of those of Tlatelulco, and of the first Kings the Mexicaines did choose.*

Sedition of Tlatelulco - - - 465
First Mexican King - - - - 466

CHAPTER 9.—*Of the strange tribute the Mexicaines paid to them of Azcapuzalco.*

Mexican tribute to Azcapuzalco - - - 468
Floating gardens - - - - - 469
Death of the first Mexican King - - - 470

CHAPTER 10.—*Of the second King, and what happened in his raigne.*

The second Mexican King - - - - 471
Death of the second King - - - - 472

CHAPTER 11.—*Of Chimalpopoca, the third king, and his cruell death, and the occasion of the warre which the Mexicaines made.*

The third Mexican King - - - - 473
Power of Mexican Kings - - - - 474
Murder of the third Mexican King - - - 475

ANALYTICAL TABLE OF CONTENTS.   xi

CHAPTER 12.—*Of the fourth King called Izcoalt, and of the war against the Tepanecas.*

The fourth Mexican King - - - - 477
The warrior Tlacaellel - - - - 478

CHAPTER 13.—*Of the battell the Mexicaines gave to the Tepanecas, and of the victory they obtained* - 480

CHAPTER 14.—*Of the warre and victory the Mexicaines had against the Cittie of Cuyoacan* - - 483

CHAPTER 15.—*Of the warre and victorie the Mexicaines had against the Suchimilcos* - - - 485

CHAPTER 16.—*Of the fift King of Mexico, called Montezuma, the first of that name* - - - 488

CHAPTER 17.—*How Tlacaellel refused to be King, and of the election and deedes of Ticocic* - - - 491

CHAPTER 18.—*Of the death of Tlacaellel, and the deedes of Axayaca, the seventh King of the Mexicaines* - - 494

CHAPTER 19.—*Of the deedes of Autzol, the eighth King of Mexico.*

Accession of the eighth king - - - 497
Conquests of Autzol - - - - 497
Machinations of a sorcerer - - - - 498
Water brought to Mexico - - - - 499

CHAPTER 20.—*Of the election of great Montezuma, the last King of Mexico.*

Character of Montezuma - - - - 500
Speech of the King of Tezcuco - - - 501
Reply of Montezuma - - - - 502

ANALYTICAL TABLE OF CONTENTS.

CHAPTER 21.—*How Montezuma ordered the service of his house, and of the warre he made for his coronation.*

Household of Montezuma - - - - 503
Coronation of Montezuma - - - 504

CHAPTER 22.—*Of the behaviour and greatnes of Montezuma.*

Government of Montezuma - - 505

CHAPTER 23.—*Of the presages and strange prodigies which happened in Mexico before the fall of their Empire.*

Signs and wonders - 506
Credibility of omens - - - 507
A talking stone - - 509
Strange omens - - - 510

CHAPTER 24.—*Of the newes Montezuma received of the Spaniards arrival in his countrey, and of the Ambassage he sent them.*

Arrival of the Spaniards - 513
Reception of Cortes - 514
Return of the ambassadors - - - 515
Terror of Montezuma - - 516

CHAPTER 25.—*Of the Spaniards entrie into Mexico.*

Montezuma's strategy - - - - 517
Meeting of Cortes and Montezuma - 518
Interview with Montezuma - 519

CHAPTER 26.—*Of the death of Montezuma and the Spaniards departure out of Mexico.*

Rising of the Mexicans - - 520
Death of Montezuma - - 522
Retreat of the Spaniards - - 522
Submission of the Mexicans - - - 523

ANALYTICAL TABLE OF CONTENTS. xiii

CHAPTER 27.—*Of some miracles which God hath showed at the Indies in favour of the faith, beyond the desert of those that wrought them.*

Santa Cruz de la Sierra -	- 524
Curing by miracles -	525
Miracle at the siege of Cuzco -	526
Divine interposition on the side of the Spaniards	526

CHAPTER 28.—*Of the manner how the Divine Providence disposed of the Indies, to give an entrie to the Christian Religion.*

Designs of Providence - -	- 527
Importance of large monarchies -	- 528
Difficulty in converting small tribes - -	- 528
Divisions among the natives a great help	529
Gallantry of the Araucans	529
Aids to conversion -	- 530
Defeat of Satan - -	- 531
Fruits of conversion - -	532
Conclusion - -	533

A Prologue to the Bookes following.

Having intreated of the Natural History of the Indies, I will hereafter discourse of the Morall History, that is to say of the deeds and customes of the Indians. For after the heaven, the temperature, the scituation, and the qualities of the new world; after the elements and mixtures—I mean mettals, plants, and beasts, whereof we have spoken in the former Bookes, as occasion did serve; both Order and Reason doth invite vs to continue and vndertake the discourse of those men which inhabite the new world. And therefore I pretend in the following bookes to speake what I thinke worthie of this subiect. And for that the intention of this Historie is not onely to give knowledge of what hath passed at the Indies, but also to continue this knowledge, to the fruite we may gather by it, which is to helpe this people for their soules health, and to glorifie the Creator and Redeemer, who hath drawne them from the obscure darkenes of their infidelitie, and imparted vnto them the admirable light of his Gospel. And therefore I will first speake in these bookes following what concernes their religion or superstition, their customes, their idolatries, and their sacrifices; and after, what concernes their policie

and government, their lawes, customes, and their deedes.
And for that the memorie is preserved amongst the Mexicaine Nation, of their beginnings, successions, warres, and other things worthie the relation; besides that which shall be handled in the sixt booke, I will make a peculiar Discourse in the seventh, shewing the disposition and forewarnings this Nation had of the new Kingdome of Christ our Lord, which should be extended in these Countries, and should conquer them to himselfe, as he hath done in all the rest of the world. The which in truth is a thing worthie of great consideration, to see how the divine providence hath appointed that the light of his word should finde a passage in the furthest boundes of the world. It is not my proiect at this time to write what the Spaniardes have done in those partes, for there are bookes enow written vpon this subiect, nor yet how the Lordes servants have laboured and profited, for that requires a new labour. I will onely content my selfe to plant this Historie and relation at the doores of the Gospel, seeing it is alreadie entered, and to make knowne the Naturall and Morall things of the Indies, to the end that Christianitie may be planted and augmented, as it is expounded at large in the bookes we have written, *De procuranda Indiorum salute.* And if any one wonder at some fashions and customes of the Indies, and wil scorne them as fooles, or abhorre them as divelish and inhumane people, let him remember that the same things, yea, worse, have beene seene amongst the Greekes and Romans, who have commanded the whole world, as we may easily vnder-

stand, not onely of our Authors, as Eusebius of Cesarea, Clement of Alexandria, and others, but also of their owne, as Plinie, Dionysius of Halicarnaus, and Plutarke: for the Prince of darkness being the head of all Infidelitie, it is no new thing to finde among Infidells, cruelties, filthines, and follies fit for such a master. And although the ancient Gentiles have farre surpassed these of the new world in valour and naturall knowledge, yet may wee observe many things in them woorthie the remembrance. But to conclude, they shew to be barbarous people, who being deprived of the supernaturall light, want likewise philosophie and natural knowledge.

THE FIFT BOOKE

Of the Naturall and Morall Historie of the Indies.

CHAP. I.—*That the Pride and Malice of the Divell hath beene the cause of Idolatrie.*

LIB. V.

THE Pride and Presumption of the Divell is so great and obstinate that alwaies hee seekes and strives to be honoured as God, and doth arrogate to himselfe all hee can, whatsoever doth appertaine to the most high God, hee ceaseth not to abuse the blinde Nations of the world vpon whom the cleere light of the holy Gospel hath not yet shone.

Iob xli.

Wee read in Iob of this prowd tyrant, who settes his eyes aloft, and amongst all the sons of pride, he is the King. The holy Scripture instructes vs plainely of his vile intentions, and his overweening treason, whereby he hath pretended to make his Throne equall vnto Gods, saying in

Isaiah xiv.

Isaiah, "Thou diddest say within thy selfe, I will mount vp to heaven and set my chaire vpon all the starres of heaven, and I will sit vpon the toppe of the Firmament, and in the sides of the North, I will ascend above the height of the cloudes, and will be like to the most high." And in

Ezek. xxviii.

Ezekiel, "Thy heart was lifted up, and thou hast said, I am God, and have set in the chaire of God in the midst of the sea." Thus doth Satan continually persist in this wicked desire to make himselfe God. And although the iust and severe chastisement of the most high hath spoiled him of all his pompe and beautie, which made him grow prowd, being intreated as his fellonie and indiscretion had deserved, as it is written by the same Prophets; yet hath _{Lib. v.} he left nothing of his wickedness and perverse practises, the which he hath made manifest by all meanes possible, like a mad dogge that bites the sword wherewith he is strucken. For as it is written, the pride of such as hate God doth alwaies increase. Hence comes the continuall and strange care which this enemie of God hath alwaies had to make him to be worshipt of men, inventing so many kinds of Idolatries, whereby he hath so long held the gretest part of the world in subiection, so as there scarce remaines any one corner for God and his people of Israel. And since _{Mart. xii.} the power of the Gospel hath vanquished and disarmed him, and that by the force of the Crosse, hee hath broken and ruined the most important and puissant places of his kingdome with the like tyrannie, hee hath begunne to assaile the barbarous people and nations farthest off, striving to maintaine amrongst them his false and lying divinitie, the which the Sonne of God had taken from him in his Church, tying him with chaines as in a cage or prison, like a furious beast, to his great confusion, and reioycing of the servants of God, as he doth signify in Iob.

But in the end, although idolatrie had beene rooted out of the best and most notable partes of the worlde, yet he hath retired himself into the most remote parts, and hath ruled in that other part of the worlde which, although it be much inferiour in nobilitie, yet is it not of less compasse. There are two causes and chiefe motives for the which the divell hath so much laboured to plant idolatry and all infidelity, so as you shall hardly finde any Nation where there is not some markes thereof. The one is this great presumption and pride, which is such, that whoso would consider how hee durst affront the very Sonne of God, and true God, in saying impudently, that he should fall downe and worship him; the which he did, although he knew not certainely that this was the very God, yet had he some _{Mat. iv.}

opinion that it was the Sonne of God. A most cruell and horrible pride to dare thus impudently affront his God. Truely wee shall not finde it very strange that hee makes himselfe to be worshipped as God by ignorant Nations, seeing hee would seeke to be worshipped by God himselfe, calling himselfe God, being an abhominable and detestable creature. The other cause and motive of idolatrie is the mortall hatred he hath conceived for ever against mankinde. For as our Saviour saith, hee hath beene a murtherer from the beginning, and holdes it as a condition and inseparable qualitie of his wickednesse. And for that hee knowes the greatest misery of man is to worship the creature for God; for this reason hee never leaves to invent all sortes of Idolatries to destroy man and make him ennemy to God. There are two mischiefes which the divell causeth in idolatry: the one, that hee denies his God, according to the text, "Thou hast left thy God who created thee"; the other is, that hee doth subiect himselfe to a thing baser than himselfe; for that all creatures are inferior to the reasonable, and the divell, although hee be superior to man in nature, yet in estate he is much inferior, seeing that man in this life is capable of Divinitie and Eternitie. By this meanes God is dishonoured, and man lost in all parts by idolatry, wherewith the divell in his pride is well content.

CHAP. II.—*Of many kindes of idolatry the Indians have used.*

Idolatry, saieth the Holy-Ghost by the Wise man, is the cause, beginning, and end of all miseries; for this cause the enemy of mankinde hath multiplied so many sortes and diversities of idolatry, as it were an infinite matter to specifie them all. Yet we may reduce idolatry to two heades, the one grounded vppon naturall things, the other vpon

BELIEF IN A SUPREME BEING. 301

things imagined and made by mans invention. The first is divided into two; for eyther the thing they worship is generall, as the Sunne, Moone, Fire, Earth, and Elements, or else it is particular, as some certayne river, fountaine, tree, or forrest, when these things are not generaly worshipped in their kindes, but onely in particular. In this first kind of idolatry they have exceeded in Peru, and they properly call it Huaca. The second kinde of idolatry, which depends on mans inventions and fictions, may likewise be divided into two sortes, one which regards onely the pure arte and invention of man, as to adore the images or statues of gold, wood, or stone, of Mercury or Pallas, which neyther are, nor ever were any thing else but the bare pictures; and the other that concernes that which really hath beene, and is in trueth the same thing, but not such as idolatry faines, as the dead, or some things proper vnto them, which men worshippe through vanitie and flatterie, so as we reduce all to foure kindes of idolatry, which the infidells vse; of all which it behooveth us to speake something.

CHAP. III.—*That the Indians have some knowledge of God.*

First, although the darknesse of infidelitie holdeth these Nations in blindenesse, yet in many thinges the light of truth and reason works somewhat in them. And they commonly acknowledge a supreame Lord and Author of all things, which they of Peru called Viracocha,[1] and gave him names of great excellence, as Pachacamac, or Pachayachachic,[2] which is the Creator of heaven and earth: and Vsapu,[3]

---

[1] See *G. de la Vega* (ii, p. 66) for the meaning of the word *Viracocha*, properly, *Uira-ccocha*.
[2] *Pachacamac*, Creator of the World. *Pachayachachic*, Teacher of the World.   [3] *Sapay*, Only.

which is admirable, and other like names. Him they did worship, as the chiefest of all, whom they did honour in beholding the heaven. The like wee see amongst them of Mexico and China, and all other infidelles. Which accordeth well with that which is saide of Saint Paul, in the Acts of the Apostles, where hee did see the Inscription of an Altare, *Ignoto Deo*—To the vnknown God. Wherevpon the Apostle tooke occasion to preach unto them, saying, " He whome you worship without knowing, him doe I preach vnto you". In like sort, those which at this day do preach the Gospel to the Indians find no great difficultie to perswade them that there is a high God and Lord over all, and that this is the Christians God and the true God. And yet it hath caused great admiration in me, that although they had this knowledge, yet had they no proper name for God. If wee shall seeke into the Indian tongue for a word to answer to this name of God, as in Latin, *Deus*, in Greeke, *Theos*, in Hebrew, *El*, in Arabike, *Alla;* but wee shall not finde any in the Cuscan or Mexicaine tongues. So as such as preach or write to the Indians vse our Spanish name *Dios*, fitting it to the accent or pronunciation of the Indian tongues, the which differ much, whereby appeares the small knowledge they had of God, seeing they cannot so much as name him, if it be not by our very name: yet in trueth they had some little knowledge, and therefore in Peru they made him a rich temple, which they called Pachacamac, which was the principall Sanctuarie of the realme. And as it hath been saide, this word of Pachacamac is, as much to say, as the Creator, yet in this temple they vsed their idolatries, worshipping the divell and figures. They likewise made sacrifices and offrings to Viracocha, which held the chiefe place amongst the worships which the Kings Yncas made. Heereof they called the Spaniards Virocochas, for that they holde opinion they are the sonnes of heaven, and divine; even as others did attribute a deitie to Paul and Barnabas, calling the one Iupiter, and the other Mercurie, so woulde they offer sacrifices vnto them, as vnto gods: and as the Barbarians of Melita (which is Malta), seeing that the viper did not hurt the Apostle, they called him God.

Acts xviii.

As it is therefore a trueth, conformable to reason, that there is a soveraigne Lorde and King of heaven, whome the Gentiles, with all their infidelities and idolatries, have not denyed, as wee see in the Philosophy of Timæus in Plato, in the Metaphisickes of Aristotle, and in the Asclepio of Tresmigister, as also in the Poesies of Homer and Virgil. Therefore the Preachers of the Gospel have no great difficultie to plant and perswade this truth of a supreame God, be the Nations of whome they preach never so barbarous and brutish. But it is hard to roote out of their mindes that there is no other God, nor any other deitie then one; and that all other things of themselves have no power, being not workeing proper to themselves, but what the great and onely God and Lord doth give and impart vnto them. To conclude, it is necessarie to perswade them by all meanes in reproving their errors, as well in that wherein they generally fail in worshipping more then one God, as in particular (which is much more), to hold for Gods, and to demand favour and helpe of those things which are not Gods, nor have any power, but what the true God their Lord and Creator hath given them.

Pla. in Tim. Arist., c. vlti. 12, metaph. Tresmegist. in Pimandro. and Asclepio.

CHAP. IV.—*Of the first kinde of Idolatrie, vpon naturall and universall things.*

Next to Viracocha, or their supreme God, that which most commonly they have and do adore amongst the Infidells is the Sunne; and, after, those things which are most remarkable in the celestiall or elementarie nature, as the moone, starres, sea, and land. The Huacas, or Oratories, which the Yncas Lords of Peru had in greatest reverence next to Viracocha and the sunne, was the thunder, which they called by three divers names, Chuquilla, Catuilla, and Intiillapa,[1] supposing it to bee a man in heaven, with a sling and a mace, and that it is in his power to cause raine, haile, thunder, and all the rest that appertaines to the region of the aire, where the cloudes engender. It was a Huaca (for so they called the Oratories) generall to all the Indians of Peru, offering vnto him many sacrifices; and in Cuzco, which is the Court and Metropolitane Cittie, they did sacrifice children vnto him, as to the Sunne. They did worship these three, Viracocha, the Sunne, and Thunder, after another maner than all the rest, as Polo[2] writes, who had made triall thereof, they did put as it were a gauntlet or glove vpon their hands when they did lift them vp to worshippe them. They did worshippe the earth, which they called Pachamama, as the Ancients did the goddesse Tellus; and the sea likewise, which they call Mamacocha, as the Ancients worshipped Thetis or Neptune. Moreover, they did worship the rainebow, which were the armes and blazons of the Ynca, with two snakes stretched out on either side. Amongst the starres they all did commonly worship that which they called Colca, and we heere the little goats.[3] They did attribute divers offices to divers starres, and those which had neede of their favour did worship them, as the shepheard did sacrifice to a star which they called vrcuchillay, which they holde to be a sheepe of divers colours, having the care to preserve their cattell. It is understood to be that which the Astronomers call Lyra. These shepheards worshippe two other starres, which walke neere vnto them, they call them Catuchillay

---

[1] *Yllapa* is thunder in Quichua. *Chuqui-ylla* was the name of the God of Thunder. *Ynti-yllapa*, the Sun's thunder.

[2] Polo de Ondegardo.   [3] The Plaiades.

MEXICAN DEITIES. 305

and vrcuchillay; and they faine them to be an ewe and a lambe. Others worshipped a starre which they called Machachuay, to which they attribute the charge and power over serpents and snakes, to keepe them from hurting of them. They ascribe power to another starre, which they called Chuquinchincay (which is as much as jaguar), over tigres, beares, and lyons, and they have generally beleeved, that of all the beasts of the earth, there is one alone in heaven like vnto them, the which hath care of their procreation and increase. And so they did observe and worship divers starres, as those which they called Chacana, Topatorca, Mamana, Mirco, Miquiquiray, and many others. So, as it seemed, they approached somewhat neere the propositions of Platoes Ideas. The Mexicaines almost in the same maner, after the supreame God, worshiped the Sunne. And therefore they called Hernando Cortez, as he hath written in a letter sent vnto the Emperour Charles the fift, Sonne of the Sunne, for his care and courage to compasse the earth. But they made their greatest adoration to an Idol called Vitzilipuztli, the which in all this region they called the most puissant, and Lord of all things; for this cause the Mexicaines built him a Temple, the greatest, the fairest, the highest, and the most sumptuous of all other. The scituation and beautie thereof may wel be coniectured by the ruines which yet remaine in the midst of the Cittie of Mexico. But heere the Mexicaines Idolatrie hath bin more pernicious and hurtfull then that of the Yncas, as wee shall see plainer heereafter, for that the greatest part of their adoration and idolatrie was employed to Idols, and not to naturall things, although they did attribute naturall effects to these Idolls, as raine, multiplication of cattell, warre, and generation, even as the Greeks and Latins have forged Idolls of Phœbus, Mercurie, Iupiter, Minerva, and of Mars. To conclude, whoso shall neerely looke into it, shall finde this manner which the Divell hath vsed to deceive the

x

Indians, to be the same wherewith hee hath deceived the Greekes and Romans, and other ancient Gentiles, giving them to vnderstand that these notable creatures, the Sunne Moone, Starres, and Elements, had power and authoritie to doe good or harme to men. And although God hath created all these things for the vse of man, yet hath man so much forgotte himselfe as to rise vp against him. Moreover, he hath imbased himselfe to creatures that are inferiour vnto himselfe, worshiping and calling vpon their workes, forsaking his Creator. As the Wise man saieth well in these wordes, "All men are vaine and abused that have not the knowledge of God, seeing they could not know him, that is, by the things that seemed good vnto them: and although they have beheld his workes, yet have they not attained to know the author and maker thereof, but they have beleeved that the fire, winde, swift aire, the course of the starres, great rivers, with Sunne and Moone, were Gods and governours of the world; and being in love with the beautie of these things, they thought they should esteeme them as Gods." It is reason they should consider how much more faire the Creator is, seeing that he is the Author of beauties and makes all things. Moreover, if they admire the power and effects of these things, thereby they may vnderstand how much more mightie hee is that gave them their being, for by the beautie and greatnes of the creatures, they may iudge what the Maker is. Hitherto are the wordes of the Booke of Wisdome, from whence we may draw a good and strong argument, to overthrow the Idolatrie of Infidells, who seeke rather to serve the creature then the Creator, as the Apostle doth iustly reprehend them. But for as much as this is not of our present subiect, and that it hath been sufficiently treated of in the Sermons written against the errors of the Indians, it shall bee sufficient now to shew that they did worship the great God, and their vaine and lying gods all of one fashion; for their maner to pray to

Viracocha, to the Sunne, the Starres, and the rest of their Idolls, was to open their hands, and to make a certaine sound with their mouthes, like people that kissed, and to aske that which every one desired in offering his sacrifices, yet was there great difference betwixt the words they vsed in speaking to the great Ticciviracocha,[1] to whom they did attribute the cheefe power and commandement over all things, and those they vsed to others, the which every one did worship privately in his house, as Gods or particular Lords, saying that they were their intercessors to this great Ticciviracocha. This maner of worship, opening the hands, and as it were kissing, hath something like to that which Iob had in horror, as fit for Idolaters, saying, " If I have kissed my hands with my mouth, beholding the Sunne when it shines, or the Moone when it is light, the which is a great iniquitie, and to deny the most great God."

Lib. v.

Iob xxxii.

CHAP. V.—*Of the Idolatry the Indians vsed to particular things.*

The Divell hath not bene contented to make these blinde Indians to worshippe the Sunne, Moone, Starres, Earth and Sea, and many other generall things in nature, but hee hath passed on further, giving them for God, and making them subiect to base and abiect things, and for the most part, filthy and infamous. No man needes to woonder at this barbarous blindnes, if hee remember what the Apostle speaketh of Wise men and Philosophers. That having knowne God, they did not glorifie him, nor give him thankes as to their God, but they were lost in their own imaginations and conceipts, and their hearts were hardened in their follies, and they have changed the glory and deity of the

Rom. i.

[1] "Aticsi-Uiracocha", according to Molina. From "Atic", a conqueror.

eternall God into shews and figures of vaine and corruptible things, as men, birds, beasts, and serpents; we know well that the Egyptians did worship the Dogge of Osiris, the Cow of Isis, and the Sheepe of Ammon; the Romans did worship the goddesse Februa, of Feavers, and the Tarpeien Goose; and Athenes the wise woman, the Cocke, and the Raven, and such other like vanities and mockeries, whereof the auntient Histories of the Gentiles are full. Men fell into this great misery, for that they would not subiect themselves to the Lawe of the true God and Creator, as Saint Athanasius dooth learnedly handle, writing against Idolatry. But it is wonderfull strange to see the excesse which hath beene at the Indies, especially in Peru; for they worshipped rivers, fountaines, the mouthes of rivers, entries of mountaines, rockes or great stones, hilles and the tops of mountains, which they call Apachitas, and they hold them for matters of great devotion. To conclude, they did worship all things in nature which seemed to them remarkable and different from the rest, as acknowledging some particular deitie.

They shewed me in Caxamalca of Nasca a little hill or great mount of sand, which was the chiefe Idoll or Huaca of the Antients. I demaunded of them what divinitie they found in it? They answered, that they did worship it for the woonder, being a very high mount of sand, in the midst of very thicke mountains of stone. Wee had neede in the cittie of Kings of great store of great wood for the melting of a Bell, and therefore they cut downe a great deformed tree, which for the greatnesse and antiquitie thereof had beene a long time the Oratorie and Huaca of the Indians. And they beleeved there was a certaine Divinity in any thing that was extraordinary and strange in his kinde, attributing the like vnto small stones and mettalls; yea, vnto rootes and fruites of the earth, as the rootes they call Papas. There is a strange kinde which they call Llallahuas, which they kissed and worshipped. They did likewise worshippe Beares, Lions, Tygres, and Snakes, to the end they should not hurt them; and such as their gods bee, such are the things they offer vnto them in their worshippe. They have vsed as they goe by the way, to cast, in the crosse wayes, on the hilles, and toppes of mountaines, which they call Apachitas,[1] olde shooes, feathers, and coca chewed, being an hearb they vse much. And when they have nothing left, they cast a stone as an offring, that they might passe freely, and have greater force, the which they say increaseth by this meanes, as it is reported in a provinciall Council of Peru. And therefore they finde in the hie wayes great heapes of stones offered, and such other things. The like follie did the Antients vse, of whome it is spoke in the Proverbs. "Like vnto him that offereth stones vnto the hill of Mercurie, such a one is hee that honoureth fooles,"[2] meaning that a man shall reape no more fruit nor profit of the second than the first, for that their God Mercury, made of stone, dooth not acknowledge any offering, neyther doth a foole any honour that is doone him. They vsed another offring no lesse absurd, pulling the haire from the eyebrowes to offer it to the Sunne, hills, Apachitas, to the winds, or to any other thing they feare. Such is the miseries that many Indians have lived in, and do to this day, whom the divell doth abuse, like very babes, with any foolish illusion whatsoever. So dooth Saint Chrysostome in one of his Homilies compare them, but the servants of God, which labour to draw them to salvation, ought not contemne these follies and childishnesse, being sufficient to plunge these poore abused creatures into eternall perdition; but they ought with good and cleere reasons to draw them from so great ignorance.

[1] Correctly "Apachecta". See *G. de la Vega*, i, p. 117.
[2] "As he that bindeth a stone in a sling, so is he that giveth honour to a fool."—*Proverbs* xxvi, v. 8.

For in trueth it is a matter woorthy of consideration, to see how they subiect themselves to such as instruct them in the true way of life. There is nothing among all the creatures more beautifull than the Sunne, which all the Gentiles did commonly worship. A discreete captaine and good christian told me that he had with a good reason perswaded the Indians that the Sunne was no god. He required the Cacique or chiefe Lord to give him an Indian that were light, to carry him a letter; which doone, he saide to the Cacique, Tell me who is Lord and chiefe, either this Indian that carries the letter, or thou that dost send him? The Cacique answered, without doubt I am, for he dooth but what I commaund him. Even so replied the Captaine, is it of the Sunne we see, and the Creator of all things. For that the Sunne is but a servant to the most high Lorde, which, by his commaundement, runnes swiftly, giving light to all nations. Thus thou seest it is against reason to yeeld that honour to the Sunne which is due to the Creator and Lord of all. The Captaine's reason pleased them all; and the Cacique with his Indians sayde it was trueth, and they were much pleased to vnderstand it.

They report of one of the Kings Yncas, a man of a subtill spirite, who, seeing that all his predecessors had worshipped the Sunne, said that hee did not take the Sunne to be God, neither could it be, for that God was a great Lord, who with great quiet and leasure performeth his workes, and that the Sunne doth never cease his course, saying that the thing which laboured so much could not seeme to be God.[1] Wherein hee spake truth. Even so, when they shew the Indians their blind errors by lively and plaine reasons, they are presently perswaded and yeelde admirably to the trueth.

---

[1] This was Huayna Ccapac. See *G. de la Vega*, ii, p. 446.

CHAP. VI.—*Of another kinde of idolatry vpon the dead.*

There is an other kinde of idolatry, very different from the rest, which the Gentiles have vsed for the deads sake whom they loved and esteemed; and it seemeth that the Wise man would give vs to vnderstand, that the beginning of idolatry proceeded thence, saying thus: "The seeking of Idolles was the beginning of fornication, and the bringing vp of them is the destruction of life. For they were not from the beginning, neither shall they continue for ever, but the vanitie and idlenesse of men hath found out this invention, therefore shall they shortly come to an end; for when a father mourned heavily for the death of his miserable sonne, he made for his consolation an Image of the dead man, and beganne to worshippe him as a god, who a little before had ended his daies like a mortall man, commanding his servants to make ceremonies and sacrifices in remembrance of him. Thus in processe of time this vngratious custome waxing strong was held for a lawe, and Images were worshipped by the commaundement of Kings and Tirantes. Then they beganne to doe the like to them that were absent, and such as they could not honour in presence, being farre off, they did worship in this sort, causing the Images of Kings to be brought whom they would worship, supplying, by this invention, their absence whom they desired to flatter. The curiositie of excellent workmen increased this Idolatrie, for these Images were made so excellent by their Art, that the ignorant were provoked to worshippe them, so as by the perfection of their Arte, pretending to content them that gave them to make, they drew Pictures and Images farre more excellent; and the common people, ledde with the shew and grace of the worke, did holde and esteeme him for a God, whome before they had honoured as a man. And this was the miserable errour of men, who sometimes

LIB. V.

Wisd. xiv, 12 to 21.

yeelding to their affection and sence, sometimes to the flatterie of their Kings, did attribute vnto stones the incommunicable name of God, worshipping them for Gods."

All this is in the booke of Wisdome, woorthy to be noted; and such as are curious in the search of Antiquities shall finde that the beginning of idolatry were these Images of the dead. I say idolatry, which is properly the worship of Idolles and Images; for that it is not certaine that this other idolatry, to worship the creatures, as the Sunne and and the hostes of heaven, or the number of Planets and Starres, whereof mention is made in the Prophets, hath beene after the idolatry of Images, although without doubt they have made idols in honour of the Sunne, the Moone, and the Earth. Returning to our Indians; they came to the height of Idolatry by the same meanes the Scripture maketh mention of: first they had a care to keepe the bodies of their Kings and Noblemen whole, from any ill scent or corruption above two hundred yeares. In this sorte were their Kings Yncas in Cusco, every one in his Chappell and Oratorie, so as the Marquis of Cañete being Viceroy, to root out Idolatry, caused three or foure of their gods to be drawne out and carried to the city of Kings, which bredde a great admiration, to see these bodies (dead so many yeares before) remaine so faire and also whole.[1] Every one of these Kings Yncas left all his treasure and revenues to entertaine the place of worshippe where his body was layed, and there were many Ministers with all his familie dedicated to his service; for no King successor did vsurpe the treasures and plate of his predecessor, but he did gather all new for himselfe, and his pallace. They were not content with this Idolatry to dead bodies, but also they made their figures and representations; and every King in his life time caused a figure to be made wherein he was represented, which they called Huauque, which signifieth

[1] See *G. de la Vega*, ii, p. 91.

brother, for that they should doe to this Image, during his life and death, as much honor and reverence as to himself. They carryed this Image to the warres, and in procession for rain or fayre weather, making sundry feastes and sacrifices vnto them. There have beene many of these Idolles in Cusco, and in that territorie, but nowe they say that this superstition of worshipping of stones hath altogether ceased, or for the most part, after they had beene discovered by the diligence of the Licentiate Polo, and the first was that of the Ynca Rocca, chief of the faction or race of Hanan Cusco. And we find that among other Nations they had in great estimation and reverence the bodies of their predecessors, and did likewise worship their Images.

CHAP. VII.—*Of Superstitions they vsed to the Dead.*

The Indians of Peru beleeved commonly that the Soules lived after this life, and that the good were in glorie and the bad in paine; so as there is little difficultie to perswade them to these articles. But they are not yet come to the knowledge of that point, that the bodies should rise with the soules. And therefore they did vse a wonderfull care, as it is saide, to preserve the bodies which they honoured after death; to this end their successors gave them garments, and made sacrifices vnto them, especially the Kings Yncas, being accompanied at their funeralls with a great number of servants and women for his service in the other life; and therefore on the day of his decease they did put to death the woman he had loved best, his servants and officers, that they might serve him in the other life.

Whenas Huayna Ccapac died (who was father to Atahualpa, at what time the Spaniards entred), they put to death aboue a thousand persons of all ages and conditions, for his service, to accompany him in the other life; after many songs and drunkennes they slew them; and these that were appointed to death, held themselves happy. They did sacrifice many things vnto them, especially yong children, and with the bloud they made a stroake on the dead mans face, from one eare to the other. This superstition and inhumanitie, to kill both men and women, to accompanie and serve the dead in the other life, hath beene followed by others, and is at this day vsed amongst some other barbarous Nations. And as Polo writes, it hath beene in a maner generall throughout all the Indies. The venerable Bede reportes, that before the Englishmen were converted to the Gospel they had the same custome, to kill men to accompany and serve the dead. It is written of a Portugall, who, being captive among the Barbarians, had beene hurt with a dart, so as he lost one eye, and as they would have sacrificed him to accompany a Nobleman that was dead, hee said vnto them that those that were in the other life would make small account of the dead if they gave him a blind man for a companion, and that it were better to give him an attendant that had both his eyes. This reason being found good by the Barbarians they let him go. Besides this superstition of sacrificing men to the dead, beeing used but to great Personages, there is another far more general and common in all the Indies, which is to set meate and drinke vpon the grave of the dead, imagining they did feede thereon: the which hath likewise beene an error amongst the Ancients, as saint Augustine writes, and therefore they gave them meate and drinke. At this day many Indian Infidells doe secretly draw their dead out of the churchyard and burie them on hilles, or vpon passages of mountains, or else in their owne houses. They have also vsed to put gold and silver in their mouth, hands, and bosome, and to apparell them with new garments, durable and well lined, vnder the herse.

They beleeve that the soules of the dead wandred vp and

downe and indure colde, thirst, hunger, and travell, and for this cause they make their anniversaries, carrying them clothes, meate, and drinke. So as the Prelates, in their Synodes, above all things, give charge to their Priests to let the Indians vnderstand, that the offerings that are set vpon the sepulchre is not to feede the dead but for the poor and ministers, and that God alone dooth feede the soules in the other life, seeing they neither eate nor drinke any corporall thing, being very needefull they should vnderstand it, lest they should convert this religious vse into a superstition of the gentiles as many doe.

CHAP. VIII.—*Of the manner of burying the dead among the Mexicaine and sundrie other Nations.*

Having reported what many nations of Peru have done with their dead, it shall not be from the purpose to make particular mention of the Mexicaines in this poynt, whose mortuaries were much solemnified and full of notable follies. It was the office of the priests and religious of Mexico (who lived there with a strange observance, as shall be said hereafter) to interre the dead and doe their obsequies. The places where they buried them was in their gardens, and in the courts of their owne houses; others carried them to the places of sacrifices which were doone in the mountaines; others burnt them, and after buryed the ashes in theyr Temples, and they buryed them all with whatsoever they had of apparel, stones, and jewells. They did put the ashes of such as were burnt into pots, and with them the jewells, stones, and earerings of the dead, how rich and pretious soever. They did sing the funerall offices like to answeres, and did often lift vp the dead bodies, dooing many ceremonies. At these mortuaries they did eate and drinke, and if it were a person of qualitie they gave apparrell to all such as came to the interrement. When any one dyed they layd him open in a chamber, vntill that all his kinsfolkes and friendes were come, who brought presents vnto the dead, and saluted him as if he were living. And if he were a King or a Lord of some towne, they offered him slaves to be put to death with him, to the end they might serve him in the other world. They likewise put to death his priest or chaplaine (for every Noble man had a priest which administred these ceremonies within his house), and then they killed him that hee might execute his office with the dead. They likewise killed his cooke, his butler, his dwarfes and deformed men, by whom he was most served; neyther did they spare the very brothers of the dead, who had most served them: for it was a greatnesse amongest the Noble men to be served by theyr brethren and the rest. Finally they put to death all of his traine for the entertaining of his house in the other world; and lest poverty should oppresse them they buried with them much wealth, as golde, silver, stones, curtins of exquisite worke, bracelets of gold, and other rich peeces. And if they burned the dead, they vsed the like with all his servants and ornaments they gave him for the other world. Then tooke they all the ashes they buryed with very great solemnity. The obsequies continued tenne dayes, with songs of plaints, and lamentations, and the priests carried away the dead with so many ceremonies, and in so great number as they coulde scarce accoumpt them. To the Captaines and Noblemen they gave trophees and marks of honour according to their enterprises and valor imployed in the warres and governements; for this effect they had armes and particular blasons. They carried these markes or blasons to the place where he desired to be buried or burnt, marching before the body, and accompanying it, as it were, in procession, where the priests and officers of the Temple went with diverse furnitures and ornaments, some casting incense, others singing, and some sounding of mournefull flutes and drummes, which did much increase the sorrow of his kinsfolkes and subjects. The priest who did the office was decked with the markes of the idoll which the noble man had represented, for all noble men did represent idolles, and carried the name of some one, and for this occasion they were esteemed and honoured. The order of knighthoode did commonly carry these forsaide markes. He that should be burnt, being brought to the place appoynted, they invironed him with wood of pine trees and all his baggage, then set they fire vnto it, increasing it still with goomie wood, vntill that all were converted into ashes, then came there foorth a Priest attired like a Divell, having mouthes vpon every ioynt of him, and many eyes of glasse, holding a great staffe with the which hee did mingle all the ashes very boldly and with so terrible a gesture, as he terrified all the assistants. Sometimes the minister had other different habites according to the qualitie of the dead. I have made this digression of obsequies and funeralls vpon the idolatry and superstition they had to the dead. It is reason to returne now to our chiefe subject and to finish this matter.

CHAP. IX.—*The fourth and last kinde of Idolatry the Indians vsed, especially the Mexicaines, to Images and Idolls.*

Although in trueth God is greatly offended with these above named Idolatries, where they woorship the creatures; yet the holy Ghost doth much more reproove and condemne another kind of idolatry, and that is of those that worship Images and figures made by the hand of men, which have nothing else in them but to be of wood, stone, or mettall, and of such forme as God hath given them. And therefore the Wiseman speaketh thus of such people, "They are miserable, whose hopes may be counted among the dead, that have called the workes of mens handes gods, as golde, silver, and the invention of the likenes of beastes, or a fruitlesse stone, which hath nothing more in it than antiquitie." And hee dooth divinely follow this proposition against this errour and follie of the Gentiles; as also the Prophets Isaiah, Jeremiah, Baruc, and King David, doe treate thereof amply. It is convenient and necessary that the ministers of Christ which doe reproove the errors of idolatry, should have a good sight, and consider well these reasons which the holy-Ghost doth so lively set downe, being all reduced into a short sentence by the Prophet Hosea, "He that hath made them was a workeman, and therefore can they be no gods, therefore the Calfe of Samaria shalbe like the Spiders webbe." Returning to our purpose, there hath beene great curiositie at the Indies in making of idolles and pictures of diverse formes and matters, which they worshipped for gods, and in Peru they called them Huacas, being commonly of fowle and deformed beasts; at the least, such as I have seene, were so. I beleeve verily that the Divel, in whose honour they made these idolles, was pleased to cause himselfe to be worshipped in these deformities, and in trueth it was found so, that the Divell spake and answered in many of these Huacas or idolls, and his priests and ministers came to those Oracles of the father of lies, and such as he is, such were his counsells and prophesies. In the provinces of New Spaine, Mexico, Tescuco, Tlascalla, Cholula, and in the neighbour countries to this realme, this kinde of idolatry hath beene more practised than in any other realme of the world. And it is a prodigious thing to heare the superstitions rehersed that they have vsed in that poynt, of the which it shall not be vnpleasant to speake something. The chiefest idoll of Mexico was, as I have sayde, Vitzilipuztli. It was an image of wood, like to a man, set vpon a stoole of the colour of azure, in a brankard or litter; at every corner was a piece of wood in forme of a Serpant's head.

The stoole signified that he was set in heaven: this idoll hadde all the forehead azure, and had a band of azure vnder the nose from one eare to another: vpon his head he had a rich plume of feathers, like to the beake of a small bird, the which was covered on the toppe with gold burnished very browne: hee had in his left hand a white target, with the figures of five pine apples made of white feathers, set in a crosse: and from above issued forth a crest of gold, and at his sides hee hadde foure dartes, which (the Mexicaines say) had beene sent from heaven to do those actes and prowesses which shall be spoken of. In his right hand he had an azured staffe, cutte in fashion of a waving snake. All these ornaments, with the rest, had their meaning, as the Mexicaines doe shew: the name of Vitziliputzli signifies the left hand of a shining feather.[1]

I will speake heereafter of the prowde Temple, the sacrifices, feasts, and ceremonies of this great idoll, being very notable things. But at this present we will only shew that this idoll, thus richly appareled and deckt, was set vpon an high Altare in a small peece or boxe, well covered with linnen clothes, iewells, feathers, and ornaments of golde, with many rundles of feathers, the fairest and most exquisite that could be found: hee had alwaies a curtine before him for the greater veneration. Ioyning to the chamber or chappell of this idoll, there was a peece of lesse worke, and not so well beautified, where there was another idoll they called Tlaloc. These two idolls were alwaies together, for that they held them as companions, and of equall power. There was another idoll in Mexico, much esteemed, which was the god of repentance, and of jubilies and pardons for their sinnes. They called this idoll Tezcatlipuca; he was made of a blacke shining stone like to Iayel,[2] being attired with some ornamental devises after their manner; it had earerings of golde and silver, and through the nether lippe a small

[1] "Siniestra de pluma relumbrante."   [2] "Azauache."

tube of cristall, in length halfe a foote: in the which they sometimes put a greene feather, and sometimes an azured, which made it resemble sometimes an emerald, and sometimes a turquois: it had the haire broided and bound vp with a haire-lace of golde burnished, at the end whereof did hang an eare of golde, with two firebrands of smoake painted therein, which did signifie the prayers of the afflicted and sinners that he heard, when they recommended themselves vnto him. Betwixt the two eares hanged a number of small herons. He had a iewell hanging at his necke, so great that it covered all his stomacke: vpon his armes bracelets of golde; at his navill a rich greene stone; and in his left hand a fanne of pretious feathers, of greene, azure, and yellow, which came forth of a looking glasse of golde, shining and well burnished, and that signified, that within this looking glasse hee sawe whatsoever was doone in the world. They called this mirror or plate of golde *Itlacheaya*, which signifies his glasse for to looke in. In his right hand he held foure dartes, which signified the chastisement hee gave vnto the wicked for their sinnes. And therefore they feared this idoll most, lest he should discover their faults and offences. At his feast they had pardon of their sinnes, which was made every foure years, as shalbe declared heereafter. They held this idoll Tezcatlipuca for the god of drought, of famine, barrennesse, and pestilence: And therefore they paynted him in another forme, being set in great maiesty vppon a stoole compassed in with a red curtin, painted and wrought with the heads and bones of dead men. In the left hand it had a target with five pines, like vnto pine apples of cotton: and in the right a little dart, with a threatening countenaunce, and the arme stretcht out, as if he would cast it; and from the target came foure dartes. It had the countenance of an angry man, and in choler, the body all painted blacke, and the head full of Quales feathers. They vsed great superstition to

GODS OF THE MEXICANS. 321

this idoll, for the feare they had of it. In Cholula, which is a commonwealth of Mexico, they worshipt a famous idoll, which was the god of marchandise, being to this day greatly given to trafficke. They called it Quetzaalcoatl.

This idoll was in a great place in a temple very high: it had about it golde, silver, jewells, very rich feathers, and habites of divers colours. It had the forme of a man, but the visage of a little bird with a red bill, and above a combe full of wartes, having rankes of teeth, and the tongue hanging out. It carried vpon the head a pointed myter of painted paper, a sithe in the hand, and many toyes of golde on the legges; with a thousand other foolish inventions, whereof all had their significations; and they worshiped it, for that he enriched whome hee pleased, as Memnon and Plutus. In trueth this name which the Cholulanos gave to their God was very fitte, although they vnderstoode it not: they called it Quetzaalcoatl, signifying colour of a rich feather, for such is the divell of covetousnesse. These barbarous people contented not themselves to have gods onely, but they had goddesses also, as the Fables of Poets have brought in, and the blind gentility of the Greekes and Romans worshipt them. The chiefe goddesse they worshipt was called Tozi, which is to say our grandmother, who, as the Histories of Mexico report, was daughter to the king of Culhuacan, who was the first they fleaed by the commaundement of Vitzliputzli, whom they sacrificed in this sort, being his sister, and then they beganne to flea men in their sacrifices, and to clothe the living with the skinnes of the sacrificed, having learned that their gods were pleased therewith, as also to pull the hearts out of them they sacrificed, which they learned of their god, who pulled out the hearts of such as he punished in Tulla, as shall be sayd in his place. One of these goddesses they worshipt had a sonne, who was a great hunter, whome they of Tlascalla afterwardes tooke for a god, and those were ennemies to the Mexicaines, by whose

Y

ayde the Spaniardes wonne Mexico. The province of Tlascalla is very fit for hunting, and the people are much given therevnto. They therfore made a great feast vnto this idoll, whom they painted of such a forme as it is not now needefull to loose any time in the description thereof. The feast they made was pleasant, and in this sort: They sounded a Trumpet at the breake of day, at the sound whereof they all assembled with their bowes, arrows, netts, and other instruments for hunting: then they went in procession with theyr idoll, being followed by a great number of people to a high mountayne, vpon the toppe whereof they had made a bower of leaves, and in the middest thereof an Altare richly deckt, where-vpon they placed the idoll. They marched with a great bruit of Trumpettes, Cornets, Flutes, and Drummes, and being come vnto the place they invironed this mountaine on all sides, putting fire to it on all partes: by meanes whereof manie beasts flew foorth, as stagges, connies, hares, foxes, and woolves, which went to the toppe flying from the fire. These hunters followed after with great cries and noyse of diverse instruments, hunting them to the top before the idoll, whither fled such a number of beastes, in so great a prease, that they leaped one vpon another, vpon the people, and vppon the Altare, wherein they tooke great delight. Then tooke they a great number of these beasts, and sacrificed them before the idoll, as stagges and other great beasts, pulling out their hearts, as they vse in the sacrifice of men, and with the like ceremony: which done, they tooke all their prey vppon their shoulders, and retired with their idoll in the same manner as they came, and entered the citty laden with all these things, very ioyfull, with great store of musicke, trumpets, and drummes, vntill they came to the Temple, where they placed their idoll with great reverence and solemnitie. They presently went to prepare their venison, wherewith they made a banquet to all the people; and after dinner they made their playes, representations, and daunces before the idoll. They had a great number of other idolles, of gods and goddesses; but the chiefe were of the Mexicaine Nation, and the neighbour people as is saide.

Chap. x.—*Of a strange manner of Idolatry practised amongst the Mexicaines.*

As we have saide that the kings Yncas of Peru caused Images to be made to their likenesse, which they called their Guacos or brothers,[1] causing them for to be honored like themselves: even so the Mexicains have done of their gods, which was in this sorte. They tooke a captive, such as they thought good; and afore they did sacrifice him vnto their idolls, they gave him the name of the idoll, to whome hee should be sacrificed, and apparelled him with the same ornaments like their idoll, saying, that he did represent the same idoll. And during the time that this representation lasted, which was for a yeere in some feasts, in others sixe monetbs, and in others lesse, they reverenced and worshipped him in the same maner as the proper idoll; and in the meane time he did eate, drincke, and was merry. When hee went through the streetes, the people came forth to worship him, and every one brought him an almes, with children and sicke folkes, that he might cure them, and bless them, suffering him to doe all things at his pleasure, onely hee was accompanied with tenne or twelve men lest he should flie. And he (to the end he might be reverenced as he passed) sometimes sounded vppon a small flute, that the people might prepare to worship him. The feast being come, and hee growne fatte, they killed him, opened him, and eat him, making a solempne sacrifice of him.

In trueth, it was a pittifull thing to consider in what sort Sathan held this people in his subiection, and doth many to

[1] *Huaca* was a sacred thing or place. *Huauque* is brother in Quichua.

this day, which commit the like cruelties and abominations, with the losse of the miserable soules and bodies of such as they offer to him, and he laughs and mockes at the follie of these poore miserable creatures, who deserve well for their offences, to be forsaken of the most high God, to the power of their adversary, whom they have chosen for their god and support. But seeing wee have spoken sufficient of the Indians idolatrie; it followes that we treate of their Religion, or rather Superstition, which they vse in their sacrifices, temples, ceremonies, and the rest.

CHAP. XI.—*How the Devill hath laboured to make himself equall vnto God, and to imitate him in his Sacrifices, Religion, and Sacraments.*

Before wee come to this point, we ought to consider one thing, which is worthie of speciall regard, the which is, how the Divell, by his pride, hath opposed himself to God; and that which God, by his wisedome, hath decreed for his honour and service, and for the good and health of man, the Divell strives to imitate and to pervert, to bee honoured, and to cause men to be damned: for as we see the great God hath Sacrifices, Priests, Sacraments, Religious Prophets, and Ministers, dedicated to his divine service and holy ceremonies, so the Divell hath his sacrifices, priests, his kinds of sacraments, his ministers appointed, his secluded and fained holinesse, with a thousand sortes of false prophets. All which will be pleasant to vnderstand, being declared in particular, and of no small fruite for him that shall remember, how the Divell is the father of lies, as the truth saieth in the Gospel; and therefore hee seekes to vsurpe to himselfe the glorie of God, and to counterfeit the light by his darknes. The Sooth-saiers of Egipt, taught by their master Sathan, laboured to do wonders, like vnto those of Moses and Aaron, to be equall vnto them. We reade in the Booke of Iudges, of that Micas, Priest of the vaine Idoll, which vsed the same ornaments which were vsed in the Tabernacle of the true God, as the Ephod, the Seraphin, and other things. There is scarce any thing instituted by Iesus Christ our Saviour in his Lawe of his Gospel, the which the Divell hath not counterfeited in some sort, and carried to his Gentiles, as may be seene in reading that which we hold for certaine, by the report of men worthie of credite, of the customes and ceremonies of the Indians, whereof we will treate in this Booke.

CHAP. XII.—*Of the Temples that were found at the Indies.*

Beginning then with their Temples, even as the great God would have a house dedicated, where his holy name might be honoured, and that it should be particularly vowed to his service; even so the Devil, by his wicked practises, perswaded Infidells to build him prowd Temples, and particular Oratories and Sanctuaries. In every Province of Peru, there was one principall Guaca,[1] or house of adoration; and besides it, there was one generall throughout all the Kingdome of the Yncas; amongst the which there hath beene two famous and notable, the one which they called Pachacamac, is foure leagues from Lima, whereat this day they see the ruines of a most ancient and great building, out of the which Francisco Pizarro and his people drew infinite treasure, of vessell and pottes of gold and silver, which they brought when they tooke the Ynca Atahualpa. There are certaine memories and discourses which say, that in this Temple the Divell did speake visibly, and gave answers by his Oracle, and that sometimes they did see a spotted snake; and it was a thing very common and

[1] *Huaca.*

approved at the Indies, that the Devill spake and answered in these false Sanctuaries, deceiving this miserable people. But where the Gospel is entred, and the Crosse of Christ planted, the father of lies is become mute, as Plutarch writes of his time "Cur cessaverit Pithias fondere oracula": and Iustine Martir treates amply of the silence which Christ imposed to devills, which spake by Idolls, as it had been before much prophecied of in the holy Scripture. The maner which the Infidel Ministers and Enchanters had to consult with their gods, was as the Devill had taught them. It was commonly in the night they entred backward to their idoll, and so went bending their bodies and head, after an vglie maner, and so they consulted with him. The answer he made, was commonly like vnto a fearefull hissing, or to a gnashing which did terrifie them; and all that he did advertise or command them, was but the way to their perdition and ruine. There are few of these Oracles found now, through the mercy of God, and great powre of Iesus Christ. There hath beene in Peru another Temple and Oratorie, most esteemed, which was in the Cittie of Cusco, where at this day is the monasterie of Santo Domingo. We may see it hath been a goodly and a stately worke by the pavement and stones of the building, which remaine to this day. This Temple was like to the Pantheon of the Romans, for that it was the house and dwelling of all the gods; for the Kings Yncas did there behold the gods of all the Nations and provinces they had conquered, every Idoll having his private place, whither they of that Province came to worship it with an excessive charge of things which they brought for his service. And thereby they supposed to keep safely in obedience those Provinces which they had conquered, holding their gods as it were in hostage. In this same house was the Punchao,[1] which was an Idoll of the Sunne, of most fine gold, wrought with great riches of

---

[1] *Punchau*, the day; hence the Sun.

stones, the which was placed to the East, with so great Art, as the sun at its rising did cast his beames thereon: and as it was of most fine mettall, his beames did reflect with such a brightnes that it seemed another Sunne. The Yncas did worship this for their God, and the Pachayacha,[1] which signifies the Creator of heaven. They say, that at the spoile of this so rich a Temple, a souldier had for his part this goodly plate of gold of the Sunne. And as play was then in request he lost it all in one night at play, whence come the proverb they have in Peru for great gamesters, saying that they play the Sunne before it riseth.[2]

Chap. xiii.—*Of the Prowd Temples at Mexico.*

The Superstitions of the Mexicaines have without comparison been greater than the rest, as well in their ceremonies as in the greatnes of their Temples, the which in old time the Spaniards called by this word Cu, which word might bee taken from the Ilanders of Santo Domingo, or of Cuba, as many other wordes that are in vse, the which are neyther from Spaine nor from any other language now vsuall among the Indians, as is Mays, Chico, Vaquiano, Chapeton, and other like. There was in Mexico, this Cu, the famous Temple of Vitziliputzli; it had a very great circuite and within a faire Court. It was built of great stones, in fashion of snakes tied one to another, and the circuite was called Coatepantli, which is a circuite of snakes; vppon the toppe of every chamber and oratorie where the Idolls were, was a fine piller wrought with small stones, blacke as iette, set in goodly order, the ground raised vp with white and red, which below gave a great light; vpon the top of the pillar were battlements very artificially made, wrought like snails,

[1] *Pacha-yachachic*, "The teacher of the universe".
[2] Mancio Serra de Leguisamo. See *G. de la Vega*, i, p. 272, and note.

supported by two Indians of stone, sitting, holding candlesticks in their hands, the which were like Croisants garnished and enriched at the ends with yellow and green feathers and long fringes of the same. Within the circuite of this court there were many chambers of religious men, and others that were appointed for the service of the Priests and Popes, for so they call the soveraigne Priests which serve the Idoll. This Court is so great and spatious, as eight or ten thousand persons did dance easily in round holding hands, the which was an vsuall custome in that Realme, although it seeme to many incredible.

There were foure gates or entries, at the East, West, North, and South, at every one of these gates beganne a faire cawsey of two or three leagues long. There was in the midst of the Lake where the Cittie of Mexico is built foure large cawseies in crosse, which did much beautifie it, vpon every portall or entery was a God or Idoll, having the visage turned to the causey right against the Temple gate of Vitziliputzli. There were thirtie steppes of thirtie fadome long, and they divided from the circuit of the court by a streete that went betwixt them; vpon the toppe of these steppes there was a walke of thirtie foote broad, all plaistered with chalke, in the midst of which walke was a Pallisado artificially made of very high trees, planted in order a fadome one from another. These trees were very bigge, and all pierced with small holes from the foote to the top, and there were roddes did runne from one tree to another, to the which were chained or tied many dead mens heads. Vpon every rod were twentie sculles, and these ranckes of sculles continue from the foote to the toppe of the tree. This Pallissado was full of dead mens sculls from one end to the other, the which was a wonderfull mournefull sight and full of horror. These were the heads of such as had beene sacrificed; for after they were dead, and had eaten the flesh, the head was delivered to the Ministers of the Temple, which tied them in this sort vntill they fell off by morcells, and then had they a care to set others in their places. Vpon the toppe of the Temple were two stones or chappells, and in them were the two Idolls which I have spoken of, Vitziliputzli and his companion Tlalot. These Chappells were carved and graven very artificially, and so high that to ascend vp to it there was a staire of stone of sixscore steppes. Before these Chambers or Chappells there was a Court of fortie foote square, in the midst whereof was a high stone of five hand breadth, poynted in fashion of a Pyramide; it was placed there for the sacrificing of men, for being laid on their backes it made their bodies to bend, and so they did open them and pull out their hearts, as I shall show heereafter. There were in the Cittie of Mexico eight or nine other Temples, the which were ioyned one to another within one great circuite and had their private staires, their courts, their chambers, and their dortoires. The entries of some were to the East, some to the West, others to the South, and some to the North. All these Temples were curiously wrought, and compassed in with divers sortes of battlements and pictures, with many figures of stones, being accompanied and fortefied with great and large spurres or platformes. They were dedicated to divers gods; but next to the Temple of Vitziliputzli was that of Tescalipuca, which was the god of penaunce and of punishments, very high and well built.

There were foure steps to ascend, on the toppe was a flat or table of sixe score foote broad, and ioyning vnto it was a hall hanged with tapistry and curtins of diverse colours and works. The doore thereof being low and large was alwayes covered with a vaile, and none but the priests might enter in. All this Temple was beutified with diverse images and pictures most curiously; for that these two Temples were as the cathedrall churches, and the rest in respect of them as parishes and hermitages; they were so spatious and had so many chambers, that there were in them places for the ministerie, colleges, schooles, and houses for priests, whereof wee will intreate heereafter. This may suffice to conceive the devills pride and the misery of this wretched nation, who with so great expence of their goods, their labour, and their lives, did thus serve their capitall enimy, who pretended nothing more than the destruction of their soules and consumption of their bodies. But yet they were well pleased, having an opinion in their so great an error that they were great and mighty gods to whome they did these services.

CHAP. XIV.—*Of the Priestes and their offices.*

We find among all the nations of the world, men specially dedicated to the service of the true God, or to the false, which serve in sacrifices, and declare vnto the people what their gods command them. Ther was in Mexico a strange curiositie vpon this point. And the devill counterfeiting the vse of the Church of God, hath placed in the order of his Priests, some greater or superiors, and some lesse, the one as Acolites, the other as Levites, and that which hath made me most to woonder, was, that the devil would vsurpe to himselfe the service of God; yea and vse the same name: for the Mexicaines in their antient tongue called their hie Priests Papes, as they should say soveraigne Bishops, as it appeares now by their Histories. The Priests of Vitzliputzli succeeded by linages of certaine quarters of the Citty, deputed for that purpose, and those of other idolls came by election, or being offered to the temple in their infancy. The dayly exercise of the Priestes was to cast incense on the idolles, which was doone foure times in the space of a naturall day. The first at breake of day, the second at noone, the third at Sunne setting, and the fourth at midnight. At midnight all the chiefe officers of the Temple did rise, and in steade of bells, they sounded a long time vpon trumpets, cornets and flutes very heavily; which being ended, he that did the office that weeke stept foorth attyred in a white roabe after the Dalmatike manner, with a censor in his hand full of coales, which he tooke from the harth burning continually before the Altare; in the other hand he had a purse full of incense, which he cast into the censor, and as he entred the place where the idoll was, he incensed it with great reverence, then tooke he a cloth, with the which he wiped the Altar and the curtins. This doone, they went all into a Chappell, and there did a certaine kinde of rigorous and austere penaunce, beating themselves, and drawing of blood, as I shall shew in the treatise of Penance which the Divell hath taught to his creatures; and heereof they never fayled at these Mattins at Midnight. None other but the Priestes might entermeddle with their sacrifices, and every one did imploy himselfe according to his dignity and degree. They did likewise preach to the people at some feastes, as I will shew when we treate thereof. They had revenues, and great offerings were made vnto them. I will speake heereafter of their vnction in Consecrating their Priestes. In Peru the Priestes were entertained of the revenues and inheritance of their God, which they called Chacaras, which were many and also verie rich.

CHAP. XV.—*Of the monastery of Virgins which the divell hath invented for his service.*

As the religious life, (whereof many servants of God have made profession in the holy Church, immitating Iesus Christ and his holy Apostles) is very pleasing in the sight of his divine maiesty, by the which his holy Name is so honoured, and his Church beutified: So the father of lies hath laboured to imitate and counterfeit him heerein; yea, as it were, hath striven with God in the observance and austere life of his ministers. There were in Peru many monasteries of Virgines (for there are no other admitted), at the least one in everie Province. In these monasteries there were two sortes of women, one antient, which they called Mamacomas,[1] for the instruction of the yoong; and the other was of yoong maidens, placed there for a certaine time, and after they were drawn foorth, either for their gods or for the Ynca. They called this house or monastery Acllaguaçi,[2] which is to say, the house of the chosen. Every monastery had his Vicar or Governour called Appopanaca,[3] who had liberty and power to choose whome he pleased, of what qualitie soever, being vnder eyght yeares of age, if they seemed to be of a good stature and constitution.

These Virgines thus shut vp into these monasteries were instructed by the Mamacomas in diverse thinges needefull for the life of man, and in the customes and ceremonies of their gods; and afterwards they tooke them from thence, being above foureteene, sending them to the Court with suregards, whereof some were appoynted to serve the Guacas and Sanctuaries, keeping their virginities for ever: some others were for the ordinary sacrifices that were made of maidens, and other extraordinary sacrifices, they made for the health, death, or warres of the Ynca: and the rest served for wives and concubines to the Ynca, and vnto other his kinsfolkes and captaines, vnto whome hee gave them, which was a great and honourable recompence: This distribution was vsed every yeare. These monasteries possessed rents and revenues for the maintenaunce of these Virgins, which were in great numbers. It was not lawfull for any father to refuse his daughters when the Appopanaca

---

[1] *Mama-cuna*, "Mothers". *Cuna* is the plural particle.

[2] *Aclla*, selected or chosen; *Huasi*, a house.

[3] *Apu*, chief; *Panaca*, from *Pana*, which means the sister of a brother. *Panaca* is the archaic genitive. *Apu-panaca* is literally "the chief over sisters of the brethren". The *Apu-panaca* was the official who selected the virgins, one over every *Hunu* or 10,000 souls.

SACRED VIRGINS IN MEXICO. 333

required them for the service of these monasteries. Yea, many fathers did willingly offer their daughters, supposing it was a great merit to be sacrificed for the Ynca. If any of these Momacomas or Acllas were found to have trespassed against their honour, it was an inevitable chasticement to bury them alive, or to put them to death by some other kind of cruell torment.

LIB. V.

The devill hath even in Mexico had some kind of religious women, although their profession was but for one yeare, and it was in this sorte: Within this great circuit whereof we have spoken, which was in the principall temple, there were two houses like cloysters, the one opposite to the other, one of men, the other of women: In that of women, they were virgines onely, of twelve or thirteene yeares of age, which they called the Maydes of Penaunce. They were as many as the men, and lived chastly and regularly, as virgins dedicated to the service of their god. Their charge was, to sweepe and make cleane the temple, and every morning to prepare meate for the idoll and his ministers, of the almes the religious gathered. The foode they prepared for the idoll were small loaves in the forme of handes and feete, and others twisted as marchpane;[1] and with this bread they prepared certaine sawses, which they cast dayly before the idoll, and his priests did eate it, as those of Baal, that Daniel speaketh of. These virgins had their haire cutte, and then they let them growe for a certaine time: they rose at midnight to the idolls mattins, which they dayly celebrated, performing the same exercises the religious did. They had their Abbesses, who imployed them to make cloth of diverse fashions for the ornament of their idolls and temples. Their ordinary habite was all white, without any worke or colour. They did their penance at midnight, sacrificing and wounding themselves, and, piercing the toppe of their eares, they layde the blood which issued foorth vpon their cheekes; and after, to wash

Dan. xiv.

[1] *Melcochas*, honey cakes.

off the blood, they bathed themselves in a pool, which was within their monastery. They lived very honestly and discreetly; and if any were found to have offended, although but lightly, presently they were put to death without remission, saying, shee had polluted the house of their god. They helde it for an augure and advertisement, that some one of the religious, man or woman, had committed a fault when they saw a Ratte or a Mowse passe, or a Bat in the chappell of their idoll, or that they had gnawed any of the vailes; for that they say a Catte or a Bat would not adventure to committe such an indignity, if some offence had not gone before, and then they beganne to make search of the fact, and having discovered the offender or offenders, of what quality soever, they presently put them to death.

None were receyved into this monastery but the daughters of one of the sixe quarters, named for that purpose: and this profession continued, as I have sayd, the space of one whole yeare: during the which time, their fathers, and they themselves, had made a vowe to serve the idoll in this manner, and from thence they went to be married. These virgins of Mexico, and more especially they of Peru, had some resemblance to the Vestall Virgins of Rome, as the Histories shew, to the end wee may vnderstand how the devill hath desired to be served by them that observe Virginitie, not that chastitie is pleasing vnto him, for he is an vncleane spirite, but for the desire he hath to take from the great God, as much as in him lieth, this glory to be served with cleannesse and integrity.

CHAP. XVI.—*Of the Monasteries of religious men that the devill hath invented for superstition.*

It is well knowne, by Letters written by the fathers of our company from Iappon, the number aud multitude of religious men that are in those Provinces, whome they call

Bonços, and also their superstitions, customes, and lies. Some fathers that have been in those countries report of these Bonços and religious men of China, saying, that there are many Orders, and of diverse sortes, some came vnto them clad in white, bearing hoodes, and others all in blacke, without haire or hoode, and these are commonly little esteemed, for the Mandarins or ministers of Iustice whippe them, as they do the rest of the people. They make profession not to eate any flesh, fish, nor any thing that hath life, but onely Rice and hearbes; but in secret they do eate any thing, and are worse than the common people. They say the religious men which are at the Court, which is at Paquin,[1] are very much esteemed. The Mandarins go commonly to recreate themselves at the Varelas[2] or monasteries of these Monkes, and returne in a manner alwayes drunke. These monasteries commonly are without the townes, and have temples within their close: yet, in China they are not greatly curious of idolles, or of temples, for the Mandarins little esteeme idolls, and do hold it for a vaine thing, and worthy to be laughed at; yea, they beleeve there is no other life, nor Paradice, but to be in the office of the Mandarins, nor any other hel than the prisons they have for offendours. As for the common sorte, they say it is necessary to entertaine them with idolatry, as the Philosopher himself teacheth his Governors: and in the Scripture it was an excuse which Aaron gave for the idol of the Calfe, that he caused to be made; yet the Chinois vsed to carry in the poupe of their shippes, in little chapels, a virgin imbosst, set in a chaire with two Chinois before her kneeling in maner of Angels, having a light burning there both day and night. And when they are to sette saile they do many sacrifices and ceremonies, with a great noyse of drummes and bells, casting papers burnt at the poupe.

Comming to our religious men, I doe not knowe that in

[1] Peking.   [2] *Viharas*.

Peru there is any proper houses for men, but for the Priests and Sorcerers, whereof there is an infinite number. But it seemeth, that in Mexico the devil hath set a due observation; for within the circuit of the great temple there were two monasteries, as before hath bin sayd, one of Virgins, whereof I have spoken, the other of yoong men secluded, of eighteene or twenty yeares of age, which they called religious. They weare shaved crownes, as the Friars in these partes, their haire a little longer which fell to the middest of their eare, except the hinder part of the head, which they let growe the breadth of foure fingers downe to their shoulders, and which they tied vppe in tresses. These young men that served in the temple of Vitzliputzli lived poorely and chastely, and did the office of Levites, ministring to the priests and chiefe of the temple their incense, lights, and garments; they swept and made cleane the holy places, bringing wood for a continual fire to the harth of their god, which was like a lampe that stille burnt before the Altar of their idoll. Besides these yong men there were other little boyes, as novices, that served for manuall vses, as to deck the temple with boughs, roses, and reeds, give the Priests water to wash with, give them their rasors to sacrifice, and goe with such as begged almes to carry it. All these had their superiors, who had the governement over them; they lived so honestly, as when they came in publike where there were any women, they carried their heads very lowe, with their eyes to the ground, not daring to beholde them; they had linnen garments, and it was lawfull for them to goe into the Citty foure or sixe together, to aske almes in all quarters: and when they gave them none, it was lawful to go into the corne fields and gather the eares of corne or clusters of mays, which they most needed, the Maister not daring to speake nor hinder them. They had this liberty because they lived poorely, and had no other revenues but almes. There might not be above fifty live in penance, rising at midnight to sound the cornets and trumpets to awake the people. Every one watched the idoll in his turne, lest the fire before the Altar should die; they gave the censor, with the which the Priest at midnight incensed the idoll, and also in the morning, at noone, and at night. They were very subject and obedient to their superiors, and passed not any one poynt that was commaunded them. And at midnight, after the priest had ended his censing, they retired themselves into a secret place apart, sacrificing and drawing blood from the calfes of their legges with sharpe bodkins; with this blood they rubbed their temples and vnder their eares; and, this sacrifice finished, they presently washt themselves in a little poole appoynted to that end. These yong men did not annoint their heads and bodies with any *betun*[1] as the Priestes did; their garments were of a coarse white linnen cloth they do make there. These exercises and strictnesse of penance continued a whole yeare, during which time they lived with great austeritie and solitarinesse. In truth it is very strange to see that this false opinion of religion hath so great force among these yoong men and maidens of Mexico that they will serve the Divell with so great rigor and austerity, which many of vs doe not in the service of the most high God, the which is a great shame and confusion; for those amongst vs that glory to have doone a small penaunce, although this exercise of the Mexicaines was not continuall, but for a yeare onely, which made it the more tollerable.

CHAP. XVII.—*Of Penance and the Strictnes the Indians have vsed at the Divell's perswasion.*

Seeing we are come to this point, it shall bee good both to discover the cursed pride of Sathan and to confound it,

[1] Pitch, a coarse wax.

and somewhat to quicken our coldnes and sloth in the service of the great God; to speake something of the rigor and strange penance this miserable people vsed at the Divell's perswasion, like to the false Prophets of Baal, who did beate and wound themselves with lancets, drawing forth bloud; or, like those that sacrificed their sonnes and daughters vnto loathsome Belphegor,[1] passing them through the fire, as holy Writ testifieth; for Sathan hath always desired to be served, to the great hurte and spoyle of man. It hath beene said that the priests and religious of Mexico rose at midnight, and having cast incense before the idoll, they retired themselves into a large place, where there were many lights; and, sitting downe, every one took a poynt of Maguay,[2] which is like vnto an awle or sharpe bodkin, with the which, or with some other kindes of launcets or rasors, they pierced the calfes of their legges neare to the bone, drawing foorth much blood, with the which they annoynted their temples, and dipt these bodkins or lancets in the rest of the blood, then set they them vpon the battlements of the Court, stickt in gloabes or bowles of strawe, that all might see and know the penance they did for the people: they do wash off the blood in a lake appoynted for that purpose, which they call Ezapangue, which is to say water of blood.

There were in the Temple a great number of bodkins or lancets, for that they might not vse one twice. Moreover, these Priests and Religious men vsed great fastings, of five or ten daies together, before any of their great feastes, and they were vnto them as our foure ember weekes; they were so strict in continence that some of them (not to fall into any sensualitie) slit their members in the midst, and did a thousand thinges to make themselves vnable, lest they should offend their gods. They drunke no wine and slept little, for that the greatest part of their exercises were by night, com-

---

[1] "Al suzio Beelfegor."
[2] Maguey, Mexican aloe.

mitting great cruelties and martiring themselves for the Divell, and all to be reputed great fasters and penitents.

They did vse to discipline themselves with cordes full of knottes, and not they onely, but the people also vsed this punishment and whipping in the procession and feast they made to the idoll Tezcatlipuca,[1] the which (as I have said before) is the god of penance; for then they all carried in their hands new cordes of the threed of Maguey a fadome long, with a knot at the end, and therewith they whipped themselves, giving great lashes over their shoulders. The Priests did fast five daies together before this feast, eating but once a day, and they lived apart from their wives, not going out of the Temple during those five daies; they did whip themselves rigorously in the manner aforesaid. The Iesuites which have written from the Indies treate amply of the penances and exceeding rigor the Bonzes[2] vse, all which was but counterfait, and more in shew then in trueth. In Peru, to solemnize the feast of the Ytu[3] which was great, all the people fasted two daies; during the which they did not accompany with their wives, neyther did they eate any meate with salt or *axi*,[4] nor drinke chicha. They did much vse this kinde of fasting for some sinnes, and did penance, whipping themselves with sharp stinging nettles, and often they strooke themselves over the shoulders with certain stones. This blinde nation, by the perswasion of the Divell, did transport themselves into craggy mountaines, where sometimes they sacrificed themselves, casting themselves downe from some high rocke. All which are but snares and deceites of him that desires nothing more then the losse and ruine of man.

[1] Tezcatlipoca was the most important of the Mexican gods. The prayers to him are given by Sahagun. His principal image was cut out of obsidian.   [2] Bonzes, Buddhist Priests.
[3] *Hatun* is "great" in Quichua.   [4] Chile pepper.

CHAP. XVIII.—*Of the Sacrifices the Indians made to the Divell, and whereof.*

It hath beene in the aboundance and diversitie of Offrings and Sacrifices taught vnto the Infidells for their idolatrie, that the enemy of God and man hath most shewed his subtiltie and wickednes. And as it is a fit thing and proper to religion to consume the substance of the creatures for the service and honour of the Creator, the which is by sacrifice, even so the father of lies hath invented the meanes to cause the creatures of God to be offered vnto him, as to the Author and Lord thereof. The first kinde of sacrifices which men vsed was very simple; for Caine offered the fruites of the earth, and Abell the best of his cattell, the which likewise Noe and Abraham did afterwardes and the other patriarkes, vntil that this ample ceremony of Levi was given by Moses, wherein there are so many sortes and differences of sacrifices of divers things for divers affaires and with divers ceremonies. In like sort, among some nations, hee hath beene content to teach them to sacrifice of what they had; but, among others, hee hath passed farre, giving them a multitude of customes and ceremonies vpon sacrifices, and so many observances as they are wonderfull. And thereby it appeares plainely that he meanes to contend and equall himselfe to the ancient law, and in many things vsurpe the same ceremonies. Wee may draw all the sacrifices the Infidells vse into three kindes—one of insensible things, another of beasts, and the third of men. They did vse in Peru to sacrifice coca which is an hearb they esteeme much, of mays which is their wheate, of coloured feathers, and of *chaquira*[1] which otherwise they call *mollo*,[2] of shelles or oysters, and sometimes gold and silver being in figures of little beasts.

---

[1] *Chaquira.* See *Cieza de Leon*, pp. 176, 405; and *G. de la Vega*, i, lib. VIII, cap. 5.

[2] *Mullu*, Quichua for a shell.

Also of the fine stuffe of *Cumbi*,[1] of carved and sweete wood, and most commonly tallow burnt. They made these offerings or sacrifices for a prosperous winde, and faire weather, or for their health, and to be delivered from some dangers and mishappes. Of the second kinde their ordinary sacrifice was of *Cuyes*,[2] which are small beasts like rabbets, the which the Indians eate commonly. And in matters of importance, or when they were rich men, they did offer *Pacos*,[3] or Indian sheepe bare or with wooll, observing curiously the numbers, colours, and times. The manner of killing their sacrifices, great or small, which the Indians did vse according to their ancient ceremonies, is the same the Moores vse at this day, the which they call *Alquible*,[4] hanging the beast by the right fore legge, turning his eyes towards the sun, speaking certain wordes according to the qualitie of the sacrifice they slew; for, if it were of colour, their words were directed to *Chuquilla*[5] and to the Thunder, that they might want no water; if it were white and smoothe they did offer it to the Sunne with certain words; if it had a fleece they did likewise offer it him with some others, that he might shine vpon them and favour their generation; if it were a *Guanaco*, which is gray, they directed their sacrifice to Viracocha. In Cusco they did every yeare kill and sacrifice with this ceremony a shorne sheepe to the Sunne, and did burne it, clad in a red waste-coate; and when they did burne it, they cast certaine small baskets of Coca into the fire, which they call *Vilcaronca*, for which sacrifice they have both men and beasts appointed which serve to no other vse. They did likewise sacrifice small birdes, although it were not so vsuall in Peru as in Mexico, where the sacrificing of

[1] *Ccompi*, Quichua for fine cloth. See *G. de la Vega*, i, lib. v, cap. 6.
[2] *Cuy* (for *Ccoy*), a guinea pig. See *G. de la Vega*, i, lib. vi, cap. 6.
[3] Alpacas.
[4] *Kibla*, the place to which Muhammadans look when praying.
[5] *Chuquilla (Chuqui-ylla)*, the Peruvian god of thunder and lightning. See Molina MS. (*Laws and Rites of the Yncas*, pp. 26, 56, 155, 167).

quailes was very ordinarie. Those of Peru did sacrifice the birdes of the Puna, for so they call the desart, when they should go to the warres, for to weaken the forces of their adversaries Huacas. They called these sacrifices *Cuzcovicsa*, or *Contevicsa*, or *Huallavicsa*, or *Sopavicsa*, and they did it in this manner: they tooke many kindes of small birdes of the desart, and gathered a great deale of a thornie wood, which they called *Yanlli*, the which being kindled they gathered together these small birdes. This assembly they called *Quiso*. Then did they cast them into the fire, about the which the officers of the sacrifice went with certaine round stones carved, whereon were painted many snakes, lions, toades, and tigres, vttering this word *Vsachum*,[1] which signifies, let the victorie be given vnto vs, with other wordes, whereby they sayed the forces of their enemies Huacas were confounded. And they drew forth certaine black sheepe, which had beene kept close some daies without meate, the which they called *Vrcu*,[2] and in killing them they spake these words: "As the hearts of these beasts be weakened, so let our enemies be weakened." And if they found in these sheep that a certaine peece of flesh behind the heart were not consumed by fasting and close keeping, they then held it for an ill augure. They brought certaine black dogs, which they call *Apurùcos*,[3] and slew them, casting them into a plaine with certaine ceremonies, causing some kinde of men to eate this flesh, the which sacrifices they did lest the Ynca should be hurt by poison; and for this cause they fasted from morning vntill the stars were vp, and then they did glut and defile themselves like to the Moores. This sacrifice was most fit for them to withstand their enemies gods; and, although at this day a great part of these customes have ceased, the wars being ended, yet remaines there some relikes by reason of the private or generall quarrels of the Indians, or the

---

[1] From *Usachuni*, I accomplish.   [2] The male animal.
[3] *Apu*, chief. *Ruccu*, old or decrepid. In Quichua *allco* is a dog.

Caciques, or in their citties. They did likewise offer and sacrifice shelles of the sea which they call *Mollo*,[1] and they offered them to the fountaines and springs, saying that these shells were daughters of the sea, the mother of all waters. They gave vnto these shells sundrie names according to the color, and also they vse them to divers ends. They vsed them in a maner in all kinde of sacrifices, and yet to this day they put beaten shells in their Chicha for a superstition. Finally they thought it convenient to offer sacrifices of everything they did sow or raise vp. There were Indians appointed to doe these sacrifices to the fountaines, springs, and rivers, which passed through the townes or by their Chacras, which are their farmes, which they did after seede time, that they might not cease running, but alwaies water their groundes. The sorcerers did coniure to know what time the sacrifices should be made, which, being ended, they did gather of the contribution of the people what should be sacrificed and delivered them to such as had the charge of these sacrifices. They made them in the beginning of winter, at such time as the fountaines, springs, and rivers did increase by the moistures of the weather, which they did attribute to their sacrifices. They did not sacrifice to the fountaines and springs of the desarts. To this day continues the respect they had to fountaines, springs, pooles, brookes, or rivers which passe by their citties or chacras, even vnto the fountaines and rivers of the desarts. They have a speciall regard and reverence to the meeting of two rivers, and there they wash themselves for their health, anointing themselves first with the flower of mays, or some other things, adding therevnto divers ceremonies, the which they do likewise in their bathes.

[1] *Mullu*, a shell.

CHAP. XIX.—*Of the Sacrifices they made of men.*

The most pittifull disaster of this poore people is their slavery vnto the Devill, sacrificing men vnto him, which are the Images of God. In many nations they had vsed to kill (to accompany the dead, as hath beene declared) such persons as had been agreeable vnto him, and whome they imagined might best serve him in the other world. Besides this, they vsed in Peru to sacrifice yong children of foure or six yeares old vnto tenne; and the greatest parte of these sacrifices were for the affaires that did import the Ynca, as in sickness for his health, and when he went to the warres for victory, or when they gave the wreathe to their new Ynca, which is the marke of a King, as heere the Scepter and the Crowne be. In this solemnitie they sacrificed the number of two hundred children, from foure to ten yeares of age, which was a cruell and inhumane spectacle. The manner of the sacrifice was to drowne them and bury them with certaine representations and ceremonies; sometimes they cutte off their heads, annointing themselves with the blood from one eare to another.

They did likewise sacrifice Virgines, some of them that were brought to the Ynca from the monasteries, as hath beene saide. In this case there was a very great and generall abuse. If any Indian qualified or of the common sorte were sicke, and that the Divine told him confidently that he should die, they did then sacrifice his owne sonne to the Sunne or to Virachoca, desiring them to be satisfied with him, and that they would not deprive the father of life. This cruelty is like to that the holy Scripture speakes of, which King Moab vsed in sacrificing his first borne sonne vpon the wall in the sight of all Israel, to whome this act seemed so mournfull as they would not presse him any further, but returned to their houses. The Holy Scripture also shewes that the like kinde of sacrifice had been in vse amongst the barbarous nations of the Cananeans, and Iebuseans, and others, whereof the booke of Wisedome speakes: "They call it peace to live in so great miseries and vexations as to sacrifice their own children, or to doe other hidden sacrifices, as to watch whole nights doing the actes of fooles, and so they keepe no cleanenesse in their life, nor in their marriages, but one through envy takes away the life of another, another takes away his wife and his contentment, and all is in confusion, blood, murther, theft, deceipt, corruption, infidelitie, seditions, periuries, mutinies, forgetfulnesse of God, pollution of soules, change of sexes and birth, inconstancie of marriages, and the disorder of adultery and filthiness; for idolatry is the sincke of all miseries." The Wise man speaketh this of those people of whome David complaines, that the people of Israel had learned those customes, even to sacrifice their sonnes and daughters to the divell, the which was never pleasing nor agreeable vnto God. For as hee is the Authour of life, and hath made all these things for the commoditie and good of man, so is hee not pleased that men should take the lives one from another; although the Lord did approve and accept the willingnesse of the faithfull patriarke Abraham, yet did hee not consent to the deede, which was to cut off the head of his sonne; wherein wee see the malice and tyranny of the divell, who would be herein as God, taking pleasure to be worshipt with the effusion of man's blood, procuring by this meanes the ruine of soule and body together for the deadly hatred he beareth to man as his cruell enemy.[1]

[1] See, on the subject of Peruvian human sacrifices, the volume on *Laws and Rites of the Yncas*, pp. 54, 58, 79, 85, 100, 166. See also my note on the subject in *G. de la Vega*, i, p. 139.

CHAP. XX.—*Of the horrible sacrifices of men which the Mexicaines vsed.*

Although they of Peru have surpassed the Mexicaines in the slaughter and sacrifice of their children (for I have not read nor vnderstood that the Mexicaines vsed any such sacrifices), yet they of Mexico have exceeded them, yea, all the nations of the worlde, in the great number of men which they had sacrificed, and in the horrible maner thereof. And to the end we may see the great miserie wherein the Divell holdes this blind Nation, I wil relate particularly the custome and inhumane maner which they have observed. First, the men they did sacrifice were taken in the warres, neyther did they vse these solemne sacrifices but of Captives: so as it seemes therein they have followed the custome of the Ancients. For as some Authors say they called the sacrifice *Victima*, for this reason, because it was of a conquered thing: they also called it *Hostia quasi ab hoste*, for that it was an offering made of their enemies, although they have applied this word to all kindes of sacrifices. In truth the Mexicaines did not sacrifice any to their idolls, but Captives, and the ordinarie warres they made was onely to have Captives for their sacrifices: and therefore when they did fight they laboured to take their enemies alive, and not to kill them, to inioy their sacrifices. And this was the reason which Moteçuma gave to the Marquis del Valle,[1] when he asked of him why being so mighty, and having conquered so many kingdomes, hee had not subdued the Province of Tlascalla, which was so neere: Moteçuma answered him that for two reasons hee had not conquered that Province, although it had beene easie if he would have vndertaken it: the one was for the exercise of the youth of

[1] The title conferred upon Hernan Cortes.

Mexico, lest they should fall into idlenes and delight: the other and the chiefe cause why he had reserved this Province was to have Captives for the sacrifices of their gods. The maner they vsed in these sacrifices was, they assembled within the palisado of dead mens sculles (as hath beene said), such as should be sacrificed, vsing a certaine ceremony at the foot of the palisado, placing a great guard about them. Presently there stept foorth a Priest, attyred with a shorte surplise full of tasselles beneath, who came from the top of the temple with an idoll made of paste, of wheate and mays mingled with hony, which had the eyes made of the graines of greene glasse, and the teeth of the graines of mays; hee descended the steppes of the temple with all the speede he could, and mounted on a great stone planted vpon a high terrasse in the midst of the court. This stone was called Quauxicalli, which is to say the stone of Eagle, whereon he mounted by a little ladder, which was in the fore part of the terrase, and descended by an other staire on the other side, still embracing his idoll. Then did he mount to the place where those were that should be sacrificed, shewing this idoll to every one in particular, saying vnto them this is your god. And having ended his shew, he descended by the other side of the staires, and all such as should die went in procession vnto the place where they should be sacrificed, where they found the Ministers ready for that office. The ordinary manner of sacrificing was to open the stomake of him that was sacrificed, and having pulled out his heart halfe alive, they tumbled the man downe the staires of the Temple, which were all imbrewed and defiled with blood. And to make it the more plaine, sixe sacrificers beeing appoynted to this dignitie, came into the place of sacrifice, foure to holde the hands and feete of him that should be sacrificed, the fift to holde his head, and the sixt to open his stomacke, and to pull out the heart of the sacrificed. They called them Chachalmua, which in our tong is as much as the ministers of holy things. It was a high dignitie, and much esteemed amongest them, wherein they did inherite and succede as in a fee simple. The minister who had the office to kill, which was the sixt amongest them, was esteemed and honoured as the soveraigne Priest and Bishop, whose name was different, according to the difference of times and solemnities. Their habites were likewise divers when they came foorth to the sacrifice, according to the diversitie of times. The name of their chiefe dignitie was Papa and Topilzin;[1] their habite and robe was a red curtain, after the Dalmatica fashion, with tasselles belowe, a crowne of rich feathers, greene, white, and yellow vpon his head, and at his eares like pendants of golde, wherein were set greene stones, and vnder the lip, vpon the middest of the beard, hee had a peece like vnto a small canon of an azured stone. These sacrificers came with their faces and handes coloured with a shining blacke. The other five had their haire much curled, and tied vp with laces of leather bound about the middest of the head: vpon their forehead they carried small roundelets of paper, painted with diverse colours, and they were attired in a Dalmatica robe of white, wroght with blacke. With this attire they represented the very figure of the Divell, so as it did strike feare and terror into all the people to see them come forth with so horrible a representation. The soveraigne priest carried a great knife in his hand of a large and sharpe flint: another priest carried a coller of wood, wrought in forme of a snake: all sixe put themselves in order, ioyning to this Piramidall stone whereof I have spoken, being directly against the doore of the Chappell of their idoll. This stone was so pointed as the man which was to be sacrificed being laid thereon vpon his backe did bend in such sort as letting the knife but fall vpon his stomacke it opened very easily in the middest. When

[1] Topiltzin, the chief sacrificial priest.

the sacrificers were thus in order they drew forth such as had beene taken in warre, which were to be sacrificed at that feast, and being accompanied with a guard of men all naked they caused them to mount vp these large staires in ranke to the place where the Ministers were prepared: and as every one of them came in their order, the six sacrificers tooke the prisoner, one by one foote another by the other, and one by one hand another by the other, casting him on his backe vpon this pointed stone, where the fift of these Ministers put the coller of wood about his necke, and the high priest opened his stomacke with the knife, with a strange dexteritie and nimblenes, pulling out his heart with his hands, the which he shewed smoaking vnto the Sunne, to whom he did offer this heate and fume of the heart, and presently he turned towardes the idoll, and did cast the heart at his face, then did they cast away the body of the sacrificed, tumbling it downe the staires of the Temple, the stone being set so neere the staires as there were not two foote space betwixt the stone and the first steppe, so as with one spurne with their foote they cast the body from the toppe to the bottome. In this sort one after one they did sacrifice all those that were appointed. Being thus slain, and their bodies cast downe, their masters, or such as had taken them, went to take them vp and carried them away: then having divided them amongest them they did eate them, celebrating their feast and solemnitie. There were ever forty or fifty at the least thus sacrificed, for that they had men very expert in taking them. The neighbour Nations did the like, imitating the Mexicaines in the customes and ceremonies of the service of their gods.

CHAP. XXI.—*Of another kind of sacrifices of men which the Mexicaines vsed.*

LIB. V.

There was an other kinde of sacrifice which they made in divers feasts, which they called Racaxipe Velitzli, which is as much as the fleaing of men.[1] They call it so for that in some feasts they tooke one or more slaves as they pleased, and after they had flead him they with that skinne apparelled a man appoynted to that end. This man went daunsing and leaping thorow all the houses and market places of the cittie, every one being forced to offer something vnto him: and if any one failed hee would strike him over the face with a corner of the skinne, defyling him with the congealed blood. This invention continued vntill the skinne did stinke: during which time, such as went gathered together much almes, which they imployed in necessary things for the service of their gods. In many of these feasts they made a challenge betwixt him that did sacrifice and him that should be sacrifyced thus: they tied the slave by one foote to a wheele of stone, giving him a sword and target in his handes to defend himselfe: then presently stept foorth hee that sacrificed him, armed with another sword and target: if he that should be sacrificed defends himselfe valiantly against the other, and resisted him, hee then remayned freed from the sacrifyce, winning the name of a famous Captaine, and so was reputed: but if hee were vanquished they then sacrifyced him on the stone wherevnto he was tyed. It was an other kinde of sacrifyce, whenas they appoynted any slave to be the representation of the idoll, saying that it was his picture. They every yeare gave one slave to the Priests, that they might never want the lively image of their idoll. At his fyrst entry into the office, after hee had beene well washed, they attyred

[1] *Xipeme* means flayed.

him with all the ornaments of the idoll, giving him the same name. Hee was that whole yeare reverenced and honoured as the idoll itselfe, and had alwayes with him twelve men for his guarde, lest hee should flie, with which guarde they suffered him to goe freely, and where hee would: and if by chaunce he fled, the chiefe of the guarde was put in his place to represent the idoll, and after to be sacrificed.

This Indian had the most honourable lodging in all the temple, where he did eate and drincke, and whither all the chiefe Ministers came to serve and honour him, carrying him meate after the manner of great personages. When hee went through the streetes of the citie hee was well accompanyed with noble men; he carried a little flute in his hand, which sometimes he sounded, to give them knowledge when he passed. Then presently the women came forth with their little children in their arms, which they presented vnto him, saluting him as god. All the rest of the people did the like: at night they put him in a strong prison or cage, lest he should flie; and when the feast came they sacrificed him, as hath beene sayde. By these and manie other meanes hath the Divell abused and entertained these poore wretches, and such was the multitude of those that had beene sacrificed by this infernall cruelty as it seems a matter incredible, for they affirme there were some dayes five thousand or more, and that there were above twenty thousand sacrifyced in diverse places. The divell to intertaine this murther of men, vsed a pleasant and strange invention, which was, when it pleased the priests of Sathan they went to their Kings, telling them how their gods died for hunger, and that they should remember them. Presently they prepared themselves, and advertised one another that their gods required meate, and therefore they should command their people to be ready to goe to the warres; and thus the people assembled, and the companies appoynted went to field, where they mustred their forces; and all their quarrell and fight was to take one another for sacrifice, striving on either side to take what captives they could, so as in these battells they laboured more to take then to kill, for that all their intention was to take men alive, to give them to their idolls to eate, for after that ·maner brought they their sacrifice vnto their gods. And wee must vnderstand that never king was crowned vntill he had subdewed some province, from the which hee brought a great number of captives for the sacrifices of their gods, so as it was an infinit thing to see what blood was spilt in the honour of the Divell.

CHAP. XXII.—*How the Indians grew weary and could not endure the cruelty of Sathan.*

Many of these Barbarians were nowe wearied and tyred with such an excessive cruelty in sheading so much blood, and with so tedious a tribute to be alwayes troubled to get captives for the feeding of their gods, seeming vnto them a matter supportable; yet left they not to followe and execute their rigorous lawes, for the great awe the ministers of these idols kept them in and the cunning wherewith they abused this poore people. But inwardly they desired to be freed from so heavy a yoke. And it was a great providence of God that the first which gave them knowledge of the Lawe of Christ found them in this disposition; for, without doubt, it seemed to them a good law and a good God to be served in this sorte. Heerevpon a grave religious man in New Spain told me that when he was in that country hee had demaunded of an auntient Indian, a man of qualitie, for what reason the Indians hadde so soone received the Lawe of Iesus Christ and left their owne, without making any other proofe, triall, or dispute thereon, for it seemed they had changed their religion without any sufficient reason to moove them. The Indian answered him: "Beleeve not, Father, that we have embraced the Law of Christ so rashly as they say, for I will tell you that we were already weary and discontented with such things as the idolls commaunded vs, and were determined to leave it and to take another Law. But whenas we found that the religion that you preached had no cruelties in it, and that it was fit for vs and both iust and good, we vnderstood and beleeved that it was the true Law, and so we received it willingly." Which answer of this Indian agrees well with that we read in the first Discourse, that Fernand Cortes sent to the Emperor Charles the Fift, wherein hee reportes that after he had conquered the city of Mexico, being in Cuyoacan, there came Ambassadors to him from the province and commonwealth of Mechoacan, requiring him to send them his law and that he would teach them to vnderstand it, because they intended to leave their owne, which seemed not good vnto them, which Cortes graunted, and at this day they are the best Indians and the truest Christians that are in New Spaine. The Spaniards that saw these cruell sacrifices resolved with all their power to abolish so detestable and cursed a butchering of men, and the rather for that in one night before their eies they sawe threescore or threescore and tenne Spaniards sacrificed, which had beene taken in a battell given at the conquest of Mexico; and another time they found written with a cole in a chamber in Tezcuco these wordes: "Here such a miserable man was prisoner with his companions whom they of Tezcuco did sacrifice."

There happened a very strange thing vpon this subiect, and yet true, being reported by men worthie of credite; which was that the Spaniards beholding these sacrifices, having opened and drawne out the heart of the lustie yong man, and cast him from the toppe of the staires (as their custome was) when hee came at the bottome, he said to the

A A

Spaniards in his language, "Knightes, they have slaine me," the which did greatly moove our men to horror and pittie. It is no incredible thing that having his heart pulled out hee might speake, seeing that Galen reports that it hath often chanced in the sacrifice of beasts, after the heart hath been drawne out and cast vpon the altar the beasts have breathed; yea, they did bray and cry out alowde, and sometimes did runne. Leaving this question how this might bee in nature, I will follow my purpose, which is to shew how much these barbarous people did now abhorre this insuportable slaverie they had to that infernall murthering, and how great the mercy of the Lord hath beene vnto them, imparting his most sweete and agreeable law.

<small>Galen., lib. ii, de Hip. and Platon. placit., cap. 4.</small>

CHAP. XXIII.—*How the Divell hath laboured to imitate and counterfaite the Sacraments of the holy Church.*

That which is most admirable in the hatred and presumption of Sathan is, that he hath not onely counterfaited in idolatry and sacrifices but also in certaine ceremonies our sacraments, which Iesus Christ our Lord hath instituted and the holy Church doth vse, having especially pretended to imitate in some sort the Sacrament of the Communion, which is the most high and divine of all others, for the great error of Infidells which proceeded in this maner. In the first moneth, which in Peru they called Rayme[1] and answereth to our December, they made a most solemne feast called Capacrayme,[2] wherein they made many sacrifices and ceremonies, which continued many daies, during the which no stranger was suffered to bee at the Court, which was in Cusco. These daies being past, they then gave libertie to strangers to enter, that they might be partakers of the feastes and sacrifices, ministring to them in this maner.

[1] *Raymi* was the month of June.
[2] *Ccapac Raymi* was the solstice of December.

The Mamaconas of the Sunne, which were a kinde of Nunnes of the Sunne, made little loaves of the flower of Mays, died and mingled with the bloud of white sheepe, which they did sacrifice that day; then presently they commanded that all strangers should enter, who set themselves in order; and the Priests, which were of a certaine lineage, discending from Liuquiyupangui,[1] gave to every one a morcell of these small loaves, saying vnto them that they gave these peeces to the end they should be vnited and confederate with the Ynca, and that they advised them not to speake nor thinke any ill against the Ynca, but alwaies to beare him good affection, for that this peece should be a witnesse of their intentions and will, and if they did not as they ought he would discover them and be against them. They carried these small loaves in great platters of gold and silver appointed for that vse, and all did receive and eate these peeces, thanking the Sunne infinitely for so great a favour which hee had done them, speaking wordes and making signes of great contentment and devotion; protesting that during their lives they would neither do nor thinke any thing against the Sunne nor the Ynca: and with this condition they received this foode of the Sunne, the which should remaine in their bodies for a witnesse of their fidelitie which they observed to the Sunne and to the Ynca their King. This maner of divelish communicating they likewise vsed in the tenth moneth called Coyarayme,[2] which was September, in the solemne feast which they called Cytua,[3] doing the like ceremonies. And besides this communion (if it be lawfull to vse this word in so divelish a matter) which they imparted to all strangers that came, they did likewise send of these loaves to all their Guacas, sanctuaries, or idolls, of the whole Realme; and at one instant they found people of all sides which came expresly to receive them, to whom they

[1] Lloque Yupanqui was the third sovereign of the Ynca dynasty.
[2] Ccoya Raymi.   [3] Festival of Situa.

said (in delivering them) that the Sunne had sent them that in signe that hee would have them all to worship and honour him, and likewise did sende them in honour of the Caciques. Some, perhappes, will hold this for a fable and a fiction; yet is it most true that, since the Ynca Yupangi (the which is hee that hath made most lawes, customes, and ceremonies, as Numa did in Rome), this maner of communion hath continued vntill that the Gospel of our Lord Iesus Christ thrust out all these superstitions, giving them the right foode of life, which vnites their soules to God. Whoso would satisfie himselfe more amply let him reade the relation which the Licentiate Polo did write to Don Ieronimo de Loaysa, Archbishop of the Cittie of Kings, where he shall finde this and many other things which he hath discovered and found out by his great dilligence.

CHAP. XXIV.—*In what maner the Divell hath laboured in Mexico to counterfaite the feast of the holy Sacrament and Communion vsed in the holy Church.*

It is a thing more worthy admiration to heare speak of the Feast and solemnitie of the Communion which the Divel himselfe, the Prince of Pride, ordayned in Mexico, the which (although it bee somewhat long) yet shall it not be from the purpose to relate, as it is written by men of credite. The Mexicaines in the moneth of Maie made their principall feast to their god Vitzilipuztli, and two daies before this feast, the Virgins whereof I have spoken (the which were shut vp and secluded in the same Temple and were as it were religious women) did mingle a quantitie of the seede of beetes with rosted Mays, and then they did mould it with honie, making an idoll of that paste in bignesse like to that of wood, putting insteede of eyes graines of greene glasse, of blue, or white; and for teeth graines of Mays set forth with all the ornament and furniture that I have said. This being finished, all the Noblemen came and brought it an exquisite and rich garment, like vnto that of the idol, wherewith they did attyre it. Being thus clad and deckt, they did set it in an azured chaire and in a litter to carry it on their shoulders. The morning of this feast being come, an houre before day all the maidens came forth attired in white with new ornaments, the which that day were called the Sisters of their god Vitzlipuztli, they came crowned with garlands of Mays rosted and parched, being like vnto azahar or the flower of orange; and about their neckes they had great chaines of the same, which went bauldricke-wise vnder their left arme. Their cheekes were died with vermillion, their armes from the elbow to the wrist were covered with red parrots' feathers. And thus attyred they tooke the idoll on their shoulders carrying it into the Court, where all the yoong men were attyred in garmentes of an artificiall red, crowned after the same maner like vnto the women. When as the maidens came forth with the idoll the yong men drew neer with much reverence, taking the litter wherein the idoll was vpon their shoulders, carrying it to the foote of the staires of the Temple, where all the people did humble themselves, laying earth vpon their heads, which was an ordinarie ceremonie which they did observe at the chiefe feast of their gods. This ceremony being ended, all the people went in procession with all the diligence and speede they could, going to a mountain, which was a league from the city of Mexico, called Chapultepec, and there they made sacrifices. Presently they went from thence with like diligence to go to a place neere vnto it which they called Atlacuyauaya, where they made their second station; and from thence they went to another burgh or village a league beyond Cuyoacan; from whence they parted, returning to the citie of Mexico, not making any other station. They went in this sort above foure leagues in three or foure houres, calling this procession Ypayna Vitzlipuztli. Being come to the foote of the staires they set downe the brancard or litter with the idoll, tying great cordes to the armes of the brancarde; then, with great observance and reverence, they did drawe vp the litter with the idoll in it to the top of the Temple, some drawing above and others helping belowe; in the meane time there was a great noise of fluites, trumpets, cornets, and drummes. They did mount it in this manner, for that the staires of the Temple were very steepe and narrow, so as they could not carry vp the litter vpon their shoulders, while they mounted vp the idoll all the people stoode in the Court with much reverence and feare. Being mounted to the top, and that they had placed it in a little lodge of roses which they held readie, presently came the yong men, which strawed many flowers of sundrie kindes, wherewith they filled the temple both within and without. This done all the Virgins came out of their convent, bringing peeces of paste compounded of beetes and rosted Mays, which was of the same paste whereof their idoll was made and compounded, and they were of the fashion of great bones. They delivered them to the yong men, who carried them vp and laide them at the idoll's feete, wherewith they filled the whole place that it could receive no more. They called these morcells of paste the flesh and bones of Vitzilipuztli. Having layed abroade these bones, presently came all the Ancients of the Temple, Priests, Levites, and all the rest of the Ministers, according to their dignities and antiquities (for heerein there was a strict order amongst them) one after another, with their vailes of diverse colours and workes, every one according to his dignity and office, having garlands vpon their heads and chaines of flowers about their neckes; after them came their gods and goddesses whom they worshipt, of diverse figures, attired in the same livery; then putting themselves in order about those morsells and peeces of paste, they vsed certaine ceremonies with singing and dauncing. By meanes whereof they were blessed and consecrated for the flesh and bones of this idoll.

This ceremony and blessing (whereby they were taken for the flesh and bones of the idoll) being ended they honoured those peeces in the same sorte as their god. Then came foorth the sacrificers, who beganne the sacrifice of men in the manner as hath beene spoken, and that day they did sacrifice a greater number than at any other time, for that it was the most solemne feast they observed. The sacrifices being ended, all the yoong men and maides came out of the temple attired as before, and being placed in order and ranke, one directly against another, they daunced by drummes, the which sounded in praise of the feast, and of the idoll which they did celebrate. To which song all the most ancient and greatest noble men did answer, daunsing about them, making a great circle, as their vse is, the yoong men and maides remayning alwayes in the middest. All the citty came to this goodly spectacle, and there was a commaundement very strictly observed throughout all the land, that the day of the feast of the idoll Vitzilipuztli they should eate no other meate but this paste, with hony, whereof the idoll was made. And this should be eaten at the point of day, and they should drincke no water nor any other thing till after noone: they held it for an ill signe, yea, for sacrilege to doe the contrary: but after the ceremonies ended, it was lawfull for them to eate any thing. During the time of this ceremony they hid the water from their litle children, admonishing all such as had the vse of reason not to drinke any water; which, if they did, the anger of God would come vpon them, and they should die, which they did observe very carefully and strictly. The ceremonies, dancing, and sacrifice ended, they went to vnclothe themselves, and the priests and superiors of the temple tooke the idoll of paste, which they spoyled of all the ornaments it had, and made many peeces, as well of the idoll itselfe as of the tronchons which were consecrated, and then they gave them to the people in maner of a communion, beginning with the greater, and continuing vnto the rest, both men, women, and little children, who received it with such teares, feare, and reverence as it was an admirable thing, saying that they did eate the flesh and bones of God, wherewith they were grieved. Such as had any sicke folkes demaunded thereof for them, and carried it with great reverence and veneration.

All such as did communicate were bound to give the tenth of this seede, whereof the idoll was made. The solemnitie of the idoll being ended an olde man of great authoritie stept vp into a high place, and with a lowde voice preached their lawe and ceremonies. Who would not wonder to see the divell so curious to seeke to be worshipped and reverenced in the same maner that Iesus Christ our God hath appoynted and also taught, and as the Holy Church hath accustomed. Hereby it is plainely verified what was propounded in the beginning, that Sathan strives all he can to vsurp and chalenge vnto himselfe the honor and service that is due to God alone, although he dooth still intermixe with it his cruelties and filthinesse, being the spirite of murther and vncleanenesse and the father of lies.

CHAP. XXV.—*Of Confessors and Confession which the Indians vsed.*

The father of lies would likewise counterfeit the sacrament of Confession, and in his idolatries seeke to be honored with ceremonies very like to the maner of Christians. In Peru they held opinion that all diseases and adversities came for the sinnes which they had committed, for remedy whereof they vsed sacrifices: moreover they confessed themselves verbally, almost in all provinces, and had Confessors appoynted by their superiors to that end, there were some sinnes reserved for the superiors. They received penaunce, yea, sometimes very sharpely, especially when the offendor was a poore man, and had nothing to give his Confessor. This office of Confessor was likewise exercised by women. The manner of these confessors sorcerers, whom they call Ychuiri or Ychuri,[1] hath beene most generall in the provinces of Collasuio.[2] They holde opinion that it is a heinous sinne to conceale any thing in confession. The Ychuyri or confessors discovered by lottes or by the view of some beast hides if anything were concealed, and punished them with many blowes with a stone vpon the shoulders, vntill they had revealed all: then after they gave him penaunce, and did sacrifice. They doe likewise vse this confession when their children, wives, husbands, or their Caciques be sicke, or in any great exploite. And when their Ynca was sicke all the provinces confessed themselves, chiefly those of the province of Collao. The Confessors were bound to hold their confessions secret, but in certain cases limited. The sinnes that they chiefly confessed was first to kill one another out of warre, then to steale, to take another man's wife, to give poison or sorcery to doe any harme; and they helde it to be a grievous sinne to be forgetfull in the reverence of their Guacas, or Oratories, not to observe the feasts, or to speake ill of the Ynca and to disobey him. They accused not themselves of any secret actes and sinnes. But, according to the report of some Priests, after the Christians came into that countrey, they accused themselves of their thoughts. The Ynca confessed himselfe to no man, but onely to the Sunne, that hee might tell them to Virachoca, and that he might forgive them. After the Ynca had been confessed, hee made a certaine bath to cleanse himselfe in a running river, saying these words: "I have told my sinnes to the Sunne, receive them O thou river, and carry them to

[1] *Ychurichuc* is a confessor, according to Arriaga, from *Ychurini*, I confess.   [2] Colla-suyu: the southern division of the Ynca Empire.

the sea, where they may never appeare more." Others that confessed vsed likewise these baths, with certaine ceremonies very like to those the Moores vse at this day, which they call *Guadoy*, and the Indians call them *Opacuna*.[1] When it chaunced that any man's children died he was held for a great sinner, saying that it was for his sinnes that the sonne died before the father; and, therefore, those to whom this had chanced, after they were confessed, they were bath'd in this bath called *Opacuna*, as is saide before. Then some deformed Indian, crookebackt and counterfet by nature, came to whippe them with certaine nettles. If the Sorcerers or Inchaunters by their lots and divinations affirmed that any sicke body should die, the sicke man makes no difficulty to kill his owne sonne, though he had no other, hoping by that meanes to escape death, saying that in his place he offered his sonne in sacrifice. And this crueltie hath beene practised in some places, even since the Christians came into that countrey. In trueth it is strange that this custome of confessing their secret sinnes hath continued so long amongest them, and to doe so strict penances, as to fast, to give apparell, gold, and silver, to remaine in the mountaines, and to receive many stripes vpon the shoulders. Our men say, that in the province of Chucuito, even at this day they meete with this plague of Confessors or Ychuris, whereas many sicke persons repaire vnto them; but now, by the grace of God, this people beginnes to see cleerely the effect and great benefite of our confession, wherevnto they come with great devotion. And partely this former custome hath been suffered by the providence of the Lord, that confession might not seeme tedious vnto them.

By this meanes the Lord is wholy glorified, and the Divell (who is a deceiver) deceived. And for that it concerneth this matter I will reporte the manner of a strange confession the Divell hath invented at Iappon, as appeares by a

---

[1] *Upa-cuna*, baths, from *Upani*, I wash.

letter that came from thence, which saith thus: "There are in Ocaca very great and high and stiep rockes, which have prickes or poynts on them, above two hundred fadome high. Amongest these rockes there is one of these pikes or poyntes so terribly high that when the Xamabusis (which be pilgrimes) doe but looke vp vnto it, they tremble and their haire stares, so fearefull and horrible is the place. Vpon the toppe of this poynt there is a great rod of yron of three fadome long, placed there by a strange devise; at the end of this rodde is a ballance tied, whereof the scales are so bigge as a man may sit in one of them. And the Goquis (which be divells in human shape) commaund these pilgrims to enter therein one after another, not leaving one of them; then, with an engine or instrument which mooveth by meanes of a wheele, they make this rodde of yron whereon the ballance is hanged to hang in the aire, one of these Xamabuzis being set in one of the scales of the ballaunce. And as that wherein the man is sette hath no counterpoise on the other side, it presently hangeth downe, and the other riseth vntill it meetes with and toucheth the rodde; then the Goquis telleth them from the rocke that they must confesse themselves of all the sinnes they have committed to their remembrance, and that with a lowde voyce to th'end that all the reste may heare him. Then presently hee beginneth to confesse, whilest some of the standers by do laugh at the sinnes they doe heare, and others sigh; and at every sinne they confesse the other scale of the ballance falles a little, vntill that having tolde all his sinnes it remaines equall with the other, wherein the sorrowfull penitent sits; then the Goquis turnes the wheele and drawes the rodde and ballance vnto him, and the Pilgrime comes foorth; then enters another, vntill all have passed. A Iapponois reported this after hee was christned, saying that he had beene in this pilgrimage, and entred the ballance seaven times, where he had confessed himselfe publikely. He saide, moreover, that if anie one did conceale any sinne the empty scale yeelded not; and if hee grew obstinate after instance made to confesse himselfe, refusing to open all his sinnes, the Goquis cast him downe from the toppe, where in an instant he is broken into a thousand peeces. Yet this Christian, who was called John, told vs that commonly the feare and terrour of this place is so great to all such as enter therein, and the danger they see with their eies to fall out of the ballance and to be broken in peeces, that seldome there is any one but discovers all his sins. This place is called by another name Sangenotocoro, that is to say, the place of Confession; wee see plainely by this discourse how the Divell hath pretended to vsurp vnto himselfe the service of God, making confession of sinnes (which the Lord hath appoynted for the remedy of man) a divellish superstition, to their great losse and perdition. He hath doone no lesse to the Heathen of Iappon than to those of the provinces of Collao in Peru.

CHAP. XXVI.—*Of the abominable unction which the Mexicaine priestes and other Nations vsed, and of their witchcraftes.*

God appoynted in the auntient Lawe the manner how they should consecrate Aaron's person and the other Priests, and in the Lawe of the Gospel wee have likewise the holy creame and vnction which they vse when they consecrate the Priestes of Christ. There was likewise in the auntient Lawe a sweete composition, which God defend should be employed in anie other thing then in the divine service. The Divel hath sought to counterfet all these things after his manner as hee hath accustomed, having to this end invented things so fowle and filthie, whereby they discover wel who is the Author. The priests of the idolles in Mexico were annoynted in this sort, they annointed the body from the foote to the head, and all the haire likewise, which hung like tresses, or a horse mane, for that they applyed this vnction wet and moyst. Their haire grew so as in time it hung downe to their hammes, so heavily that it was troublesome for them to beare it, for they did never cut it untill they died, or that they were dispensed with for their great age, or being employed in governments or some honorable charge in the commonwealth. They carried their haire in tresses, of sixe fingers breadth, which they died blacke with the fume of sapine, or firre trees, or rosine; for in all Antiquitie it hath bin an offring they made vnto their idolls, and for this cause it was much esteemed and reverenced. They were alwayes died with this tincture from the foote to the head, so as they were like vnto shining Negroes, and that was their ordinary vnction; yet, whenas they went to sacrifice and give incense in the mountaines, or on the tops thereof, or in any darke and obscure caves where their idolles were, they vsed an other kinde of vnction very different, doing certaine ceremonies to take away feare, and to give them courage. This vnction was made with diverse little venomous beastes, as spiders, scorpions, palmers, salamanders, and vipers, the which the boyes in the Colledges tooke and gathered together, wherein they were so expert, as they were alwayes furnished when the Priestes called for them. The chiefe care of these boyes was to hunt after these beasts; if they went any other way and by chaunce met with any of these beasts they stayed to take them, with as great paine as if their lives depended thereon. By the reason whereof the Indians commonly feared not these venomous beasts, making no more accompt than if they were not so, having beene all bred in this exercise. To make an ointment of these beastes they tooke them all together, and burnt them vpon the harth of the Temple, which was before the Altare, vntill they were consumed to ashes; then did they put them in morters with much Tobacco or *betum* (being an hearbe that Nation vseth much to benumme the flesh that they may not feele their travell), with the which they mingle the ashes, making them loose their force; they did likewise mingle with these ashes scorpions, spiders, and palmers alive, mingling all together; then did they put to it a certaine seede being grownd, which they call *Ololuchqui*, whereof the Indians make a drinke to see visions, for that the vertue of this hearbe is to deprive man of sence. They did likewise grinde with these ashes blacke and hairie wormes, whose haire only is venomous, all which they mingled together with blacke, or the fume of rosine, putting it in small pots which they set before their god, saying it was his meate: and, therefore, they called it a divine meate. By means of this oyntment they became witches, and did see and speake with the Divell. The priestes being slubbered with this oyntment lost all feare, putting on a spirit of cruelty. By reason whereof they did very boldely kill men in their sacrifices, going all alone in the night to the mountaines and into obscure caves, contemning all wilde beasts, and holding it for certayne and approved that both lions, tigres, serpents, and other furious beasts which breede in the mountaines and forrests fled from them by the vertue of this *betum* of their god.

And in trueth, though this *betum* had no power to make them flie, yet was the Divelle's picture sufficient whereinto they were transformed. This *betum* did also serve to cure the sicke and for children, and therefore all called it the Divine Physicke; and so they came from all partes to the superiors and priests, as to their saviors, that they might apply this divine physicke, wherewith they anoynted those parts that were grieved. They said that they felt heereby a notable ease, which might be, for that Tobacco and Ololuchqui have this propertie of themselves to benumme the flesh, being applied in manner of an emplaister, which must be by a stronger reason being mingled with poysons; and for that it did appease and benumme the paine, they helde it for an effect of health, and a divine virtue. And therefore ranne they to these priests as to holy men, who kept the blind and ignorant in this error, perswading them what they pleased, and making them runne after their inventions and divellish ceremonies, their authority being such as their wordes were sufficient to induce beliefe as an article of their faith. And thus made they a thousand superstitions among the vulgar people, in their maner of offering incense, in cuting their haire, tying small flowers about their necks, and strings with small bones of snakes, commaunding them to bathe at a certain time; and that they should watch all night at the harth lest the fire should die; that they should eate no other breade but that which had bin offered to their gods, that they should vpon any occasion repaire vnto their witches, who with certaine graines tolde fortunes, and divined, looking into keelers and pailes full of water. The sorcerers and ministers of the divell vsed much to besmere themselves. There were an infinite number of these witches, divines, enchanters, and other false prophets. There remaines yet at this day of this infection, althogh they be secret, not daring publikely to exercise their sacrileges, divelish ceremonies, and superstitions, but their abuses and wickednes are discovered more at large and particularly in the confessions made by the Prelates of Peru.

There is a kinde of sorcerers amongst the Indians allowed by the Kings Yncas, which are, as it were, sooth-saiers, they take vpon them what forme and figure they please, flying farre through the aire in a short time, beholding all that was done. They talke with the Divell, who answereth them in certaine stones or other things which they reverence much. They serve as coniurers, to tell what hath passed in the farthest partes, before any newes can come. As it hath chanced since the Spaniardes arrived there, that in the distance of two or three hundred leagues, they have knowne the mutinies, battailes, rebellions, and deaths, both of tyrants, and those of the King's partie, and of private men, the which have beene knowne the same day they chanced, or the day after, a thing impossible by the course of nature. To worke this divination, they shut themselves into a house, and became drunk vntil they lost their sences, a day after they answered to that which was demanded. Some affirme they vse certaine vnctions. The Indians say that the old women do commonly vse this office of witchcraft, and specially those of one Province, which they call Coaillo, and of another towne called Manchay, and of the Province of Huarochiri. They likewise shew what is become of things stolne and lost. There are of these kindes of Sorcerers in all partes, to whom commonly doe come the Anaconas,[1] and Chinas, which serve the Spaniardes, and when they have lost any thing of their masters, or when they desire to know the successe of things past or to come, as when they goe to the Spaniardes citties for their private affaires, or for the publike, they demaund if their voyage shall be prosperous, if they shall be sicke, if they shall die, or return safe, if they shall obtaine that which they pretend: and the witches or coniurers answer, yea, or no, having first spoken with the Divell, in an obscure place; so as these Anaconas do well heare the sound of the voyce, but they see not to whom these coniurers speake, neither do they vnderstand what they say. They make a thousand ceremonies and sacrifices to this effect, with the which they mocke the Divell and grow exceeding drunke, for the doing whereof, they particularly vse an hearbe called Villca,[2] the iuyce whereof they mingle with their Chicha, or take it in some other sort, whereby we may see how miserable they are, that have for their masters, the ministers of him whose office is to deceive. It is manifest that nothing doth so much let the Indians from receiving the faith of the holy Gospel, and to persever therein, as

---

[1] *Yana-cunas*, or Indians held to domestic service. See *Balboa*, p. 120. [2] A tree, the fruit of which is a purgative.—*Mossi.*

the conference with these witches, whereof there have bin, and are still, great numbers, although by the grace of the Lord, and diligence of the Prelates and Priestes, they decrease, and are not so hurtefull. Some of them have beene converted and preached publikely, discovering and blaming themselves, their errors and deceites, and manifesting their devises and lies, whereof wee have seene great effects; as also we vnderstand by letters from Jappon, that the like hath arrived in those parts: all to the glory and honour of our Lord God.

CHAP. XXVII.—*Of other Ceremonies and Customes of the Indians which are like vnto ours.*

The Indians had an infinite number of other ceremonies and customes which resembled to the ancient law of Moses, and some to those which the Moores vse, and some approached neere to the law of the Gospel, as their bathes or *Opacuna*, as they call them; they did wash themselves in water, to clense them from their sins. The Mexicaines had also amongst them a kind of baptisme, the which they did with ceremony, cutting the eares and members of yong children new borne, counterfaiting in some sort the circumcision of the Iewes. This ceremony was done principally to the sonnes of Kings and Noblemen; presently vpon their birth the priestes did wash them, and did put a little sword in the right hand, and in the left a target. And to the children of the vulgar sort they put the markes of their offices, and to their daughters instruments to spinne, knit, and labour. This ceremony continued four daies, being made before some idoll. They contracted marriage after their maner, whereof the Licentiate Polo hath written a whole Treatise, and I will speake somewhat thereon heereafter. In other things their customes and ceremonies have some show of reason. The Mexicaines were married by the handes of their priestes in this sort. The Bridegroome and the Bride stood together before the priest, who tooke them by the hands asking them if they would marrie, then having vnderstood their willes, hee tooke a corner of the vaile wherewith the woman had her head covered, and a corner of the mans gowne, the which he tied together on a knot, and so led them thus tied to the Bridegroomes house, where there was a harth kindled, and then he caused the wife to go seven times about the harth, and so the married couple sate downe together, and thus was the marriage contracted. The Mexicaines were very iealous of the integritie of their wives; so as if they found they were not as they ought to be (the which they knew eyther by signes or dishonest wordes), they presently gave notice thereof to their fathers and kinsfolkes of their wives, to their great shame and dishonor, for that they had not kept good guarde over them. But they did much honour and respect such as lived chastely, making them great banquttes, and giving great presentes both to her and to her kinsfolkes. For this occasion they made great offerings to their gods, and a solemne banket in the house of the wife, and another in the husbands. When they went to house they made an inventory of all the man and wife brought together, of provisions for the house, of land, of iewells and ornaments, which inventories every father kept, for if it chanced they made any devorce (as it was common amongest them when they agree not), they divided their goods according to the portion that every one brought, every one having libertie in such a case to marry whome they pleased; and they gave the daughters to the wife, and the sonnes to the husband. t was defended vpon paine of death, not to marry againe together, the which they observed very strictly. And although it seeme that many of their ceremonies agree with ours, yet differ they much for the great abomination they mingle therewithall.

It is common and generall to have vsually one of these three things, either cruelty, filthines, or slouth; for all their ceremonies were cruell and hurtefull, as to kill men and to spill blood, are filthy and beastly, as to eate and drinke to the name of their Idolls, and also to pisse in the honour of them, carrying them vpon their shoulders, to annoint and besmeere themselves filthily, and to do a thousand sortes of villanies, which were at the least, vaine, ridiculous, and idle, and more like the actions of children then of men. The cause thereof is the very condition of this wicked spirit, whose intention is alwaies to do ill, provoking men still to murthers and filthines, or at the least to vanities and fruitelesse actions, the which every man may well know, if he duly consider the behaviour and actions of the Divell, towardes those he sets to deceive. For in all his illusions we finde a mixture of these three, or at least of one of them. The Indians themselves (since they came to the knowledge of our faith) laugh and mocke at these fooleries and toyes, in the which their gods held them busied, whom they served more for feare, least they should hurte them, in not obeying them in all things, then for any love they bare them. Although some, yea, very many lived, abused and deceived, with the vaine hope of temporall goods, for of the eternall they had no knowledge. And whereas the temporall power was greatest, there superstition hath most increased, as we see in the Realmes of Mexico and Cusco, where it is incredible to see the number of idolls they had; for within the citty of Mexico there were above three hundred. Mango Ynca Yupangui, amongst the Kings of Cusco, was hee that most augmented the service of their idolls, inventing a thousand kindes of sacrifices, feasts, and ceremonies. The like did King Iscoalt[1] in Mexico, who was the fourth king. There was also a great number of super-

[1] Izcohuatl. He built the famous temple of Huitzilopochtli, the first god of the Mexicans.

stitions and sacrifices in other Nations of the Indians, as in the Province of Guatimala, at the Ilands in the new Kingdome, in the Province of Chile, and others that were like Commonwealthes and Comminalties. But it was nothing in respect of Mexico and Cusco, where Sathan was as in Rome, or in his Ierusalem, vntill he was cast out against his will, and the holy Crosse planted in his place, and the Kingdome of Christ our God occupied, the which the tyrant did vsurpe.

CHAP. XXVIII.—*Of some Feasts celebrated by them of Cusco, and how the Divell would imitate the mysterie of the holy Trinitie.*

To conclude that which concernes Religion, there restes something to speake of the feasts and solemnities which the Indians did celebrate, the which (for that they are divers and many) cannot be all specified. The Yncas, Lords of Peru, had two kindes of feasts, some were ordinarie, which fell out in certaine moneths of the yeere; and others extraordinary, which were for certaine causes of importance, as when they did crowne a new King, when they beganne some warre of importance, when they had any great neede of water or drought, or other like things. For the ordinary feasts, we must vnderstand, that every moneth of the yeare they made feasts, and divers sacrifices, and although all of them had this alike, that they offered a hundred sheepe, yet in colour and in forme they are very divers. In the first moneth, which they call Rayme, which is the moneth of December, they made their first feast, which was the principall of all others, and for that cause they called it Capacrayme, which is to say, a rich and principall feast. In this feast they offered a great number of sheepe and lambs in sacrifice, and they burnt them with sweete wood, then they caused gold and silver to be brought vpon certaine sheepe, setting vppon them three Images of the Sun, and three of the thunder, the father, the sonne, and the brother. In these feasts they dedicated the Yncas children, putting the Guaras or ensignes vpon them, and they pierced their eares; then some olde man did whip them with slings, and annoynted their faces with blood, all in signe that they should be true Knights to the Ynca. No stranger might remaine in Cusco during this moneth, and this feast, but at the end thereof they entred, and they gave vnto them peeces of the paste of mays with the blood of the sacrifice, which they did eate in signe of confederation with the Ynca, as hath bin said before. It is strange that the Divell after his manner hath brought a trinitie into idolatry, for the three images of the Sunne called Apomti, Churunti, and Intiquaoqui,[1] which signifieth father and lord Sunne, the sonne Sunne, and the brother Sunne. In the like maner they named the three Images of Chuquilla, which is the God that rules in the region of the aire, where it thunders, raines, and snows. I remember that, being in Chuquisaca, an honourable priest shewed me an information, which I had long in my handes, where it was prooved that there was a certaine *Huaca* or Oratory, whereas the Indians did worship an idoll called Tangatanga, which they saide was one in three, and three in one. And as this Priest stood amazed thereat, I saide that the Divell by his infernall and obstinate pride (whereby he alwayes pretendes to make himselfe God) did steale all that he could from the trueth, to imploy it in his lyings and deceits. Comming then to the feast of the second moneth, which they called Camay,[2] besides the sacrifices which they made, they did cast the ashes into the river, following five or six leagues after, praying it to carry them

---

[1] *Apu-ynti*, Chief Sun; *Churi-ynti*, Son-Sun; *Ynti-huauque*, Brother-Sun.

[2] *Canay-quilla*. The month from 8th December to 9th January.

into the sea, for that the Virochoca should there receive this present. In the third, fourth, and fift moneth, they offered a hundred blacke sheepe, speckled, and grey, with many other things, which I omitte for being too tedious. The sixt moneth is called *Hatuncuzqui Aymuray*, which answereth to Maie, in the which they sacrificed a hundred sheepe more, of all colours; in this moon and moneth, which is when they bring maize from the fieldes into the house, they made a feast, which is yet very vsuall among the Indians, and they doe call it Aymuray.[1]

This feast is made comming from the *Chacra* or farme vnto the house, saying certaine songs, and praying that the Mays may long continue, the which they call *Mamacora*. They take a certaine portion of the most fruitefull of the Mays that growes in their farmes, the which they put in a certaine granary which they doe call *Pirua*, with certaine ceremonies, watching three nightes; they put this Mays in the richest garments they have, and beeing thus wrapped and dressed, they worship this *Pirua*, and hold it in great veneration, saying it is the mother of the mays of their inheritances, and that by this means the mays augments and is preserved. In this moneth they make a particular sacrifice, and the witches demaund of this *Pirua*, if it hath strength sufficient to continue vntill the next yeare; and if it answers no, then they carry this Mays to the farme to burne, whence they brought it, according to every man's power; then make they another *Pirua*, with the same ceremonies, saying that they renue it, to the end the feede of Mays may not perish, and if it answers that it hath force sufficient to last longer they leave it vntill the next yeare. This foolish vanitie continueth to this day, and it is very common amongest the Indians to have these *Piruas*, and to make the feast of *Aymuray*. The seaventh moneth answereth to Iune, and is called *Aucaycuzqui Intiraymi*;[2] in it they

---

[1] *Aymuray*, from the middle of May. [2] *Yntip Raymi*.

made the feast that is called *Intiraymi*, in the which they sacrificed a hundred sheepe called Guanacos, and saide it was the feast of the Sunne. In this moneth they made many Images of Quinua[1] wood carved, all attired with rich garmentes, and they made their dancings which they do call *Cayo*: At this feast they cast flowers in the high wayes, and thither the Indians came painted, and their noblemen had small plates of golde vpon their beards, and all did sing; wee must vnderstand that this feast falleth almost at the same time whenas the Christians observe the solempnitie of the holy Sacrament, which doth resemble it in some sort, as in dauncing, singing, and representations. And for this cause there hath beene, and is yet among the Indians, which celebrated a feast somewhat like to ours of the holy Sacrament, many superstitions in celebrating this ancient feast of Intiraymi. The eight month is called *Chahua Huarqui*,[2] in the which they did burne a hundred sheepe more, all grey, of the colour of Viscacha, according to the former order, which month doth answer to our Iuly. The ninth moneth was called *Yapaquis*,[3] in the which they burnt an hundred sheepe more, of a chesnut colour; and they do likewise kill and burne a thousand Cuyes,[4] to the end the frost, the ayre, the water, nor the sunne should not hurt their farmes: and this moneth doth answer vnto August. The tenth moneth was called *Coyarami*,[5] in the which they burnt a hundred white sheepe that had fleeces. In this month, which answereth to September, they made the feast called *Situa* in this manner: they assembled together the first day of the moone before the rising thereof, and in seeing it they cryed aloude, carrying torches in their handes and saying, "Let all harme goe away," striking one an

---

[1] *Queñuar (Polylepis)*.
[2] The next month was *Anta-asitua* according to other authors.
[3] *Ccapac-asitua*.   [4] Guinea pigs.
[5] *Umu-Raymi* of Molina and Velasco.

other with their torches. They that did this were called *Panconcos*,[1] which being doone, they went to the common bath, to the rivers and fountaines, and every one to his own bath, setting themselves to drink foure dayes together. In this moneth the Mama-cunas of the sunne made a great number of small loaves with the blood of the sacrifices, and gave a peece to every stranger; yea, they sent to every Huaca throughout the realme, and to many Curacas, in signe of confederation and loyaltie to the Sunne and the Ynca, as hath bin said.

The bathes, drunkennesse, and some relickes of this feast Situa, remaine even vnto this day, in some places, with the ceremonies a little different, but yet very secretly, for that these chiefe and principall feasts have ceased. The eleventh moneth, *Homaraymi Punchaiquis*,[2] wherein they sacrificed a hundred sheepe more. And if they wanted water, to procure raine they set a black sheepe tied in the middest of a plaine, powring much chica about it, and giving it nothing to eate vntill it rained, which is practised at this day in many places in the time of our October. The twelfth and last month was called *Ayamarca*, wherein they did likewise sacrifice a hundred sheepe, and made the feast called *Raymicantara Rayquis*. In this moneth, which aunswered to our November, they prepared what was necessary for the children that should be made novices the moneth following; the children with the old men made a certaine shew, with rounds and turnings, and this feast was called *Ituraymi*, which commonly they make when it raines too much, or too little, or when there is a plague. Among the extraordinary feasts, which were very many, the most famous was that which they called Ytu. This feast Ytu hath no prefixed time nor season, but in time of necessitie. To prepare themselves thereunto, all the people fasted two dayes, during

---

[1] *Pancuncu*, a torch. See *G. de la Vega*, ii, p. 232.
[2] Not given by other authorities.

the which they did neyther company with their wives, nor eate anie meate with salt or garlicke, nor drinke any Chicha. All did assemble together in one place, where no straunger was admitted, nor any beast; they had garments and ornaments, which served onely for this feast. They marched very quietly in procession, their heades covered with their vailes, sounding of drummes, without speaking one to another. This continued a day and a night; then the day following they daunced and made good cheere for two dayes and two nights together, saying that their prayer was accepted. And although that this feast is not vsed at this day, with all this antient ceremony, yet commonly they make another which is verie like, which they call Ayma, with garmentes that serve onely to that end; and they make this kind of procession with their Drummes, having fasted before, then after they make good cheere, which they vsually doe in their vrgent necessities. And although the Indians forbeare to sacrifice beasts, or other things publikely, which cannot be hidden from the Spaniardes, yet doe they still vse many ceremonies that have their beginnings from these feasts and auntient superstitions; for, at this day, they do covertly make this feast of Ytu, at the dances of the feast of the Sacrament, in making the daunces of *Llama-llama*, and of *Guacon*, and of others, according to their auntient ceremonies, wherevnto we ought to take good regarde. They have made more large Discourses of that which concerneth this matter, for the necessary observation of the abuses and superstitions the Indians had in the time of their gentility, to the end the Priestes and Curates may the better take heede. Let this suffice now to have treated of the exercise wherewith the divell held those superstitious nations occupied to the end that against his will wee may see the difference there is betwixt light and darknes, betwixt the trueth of Christ and the lies of the Gentiles, although the ennemy of God and man hath laboured with all his devises to counterfet those things which are of God.

CHAP. XXIX.—*Of the feast of Iubilee which the Mexicaines celebrated.*

The Mexicaines have beene no less curious in their feasts and solemnities, which were of small charge, but of great effusion of man's blood. Wee have before spoken of the principall feast of Vitzilipuztli, after the which the feast of Tezcatlipuca was most solempnized. This feast fell in Maie, and in their Kalendar they called it Tozcoalt; it fell every foure yeeres with the feast of Penaunce, where there was given full indulgence and remission of sinnes. In this day they did sacrifice a captive which resembled the idoll Tezcatlipuca, it was the nineteenth day of Maie; upon the even of this feast the Noblemen came to the temple, bringing a new garment like vnto that of the idoll, the which the priest put vpon him, having first taken off his other garments, which they kept with as much or more reverence than we doe our ornaments. There were in the coffers of the idoll many ornaments, iewelles, eareings, and other riches, as bracelets and pretious feathers, which served to no other vse but to be there, and was worshipped as their god it selfe. Besides the garment wherewith they worshipped the idoll that day, they put vpon him certaine ensignes of feathers, with fannes, shadowes, and other things; being thus attired and furnished, they drew the curtaine or vaile from before the doore, to the ende he might be seene of all men; then came forth one of the chiefe of the temple, attired like to the idoll, carrying flowers in his hand, and a flute of earth, having a very sharpe sound, and turning towards the east, he sounded it, and then looking to the west, north, and south, he did the like. And after he had thus sounded towards the foure parts of the world (showing that both they that were present and absent did heare him) hee put his finger into the aire, and then gathered vp earth, which he put in his mouth, and did eate it in signe of adoration. The like did all they that were present, and, weeping, they fell flat to the ground, invocating the darknesse of the night, and the windes, intreating them not to leave them, nor to forget them, or else to take away their lives, and free them from the labours they indured therein. Theeves, adulterers, and murtherers, and all others offendors, had great feare and heaviness whilest this flute sounded, so as some could not dissemble nor hide their offences. By this meanes they all demanded no other thing of their god, but to have their offences concealed, powring foorth many teares, with great repentaunce and sorrow, offering great store of incense to appease their gods. The couragious and valiant men, and all the olde souldiers that followed the Arte of Warre hearing this flute, demaunded with great devotion of God the Creator, of the Lorde for whome wee live, of the sunne, and of other their gods, that they would give them victorie against their ennemies, and strength to take many captives, therewith to honour their sacrifices. This ceremonie was doone ten dayes before the feast; During which tenne dayes the Priest did sound this flute, to the end that all might do this worship in eating of earth, and demaund of their idol what they pleased: they every day made their praiers, with their eyes lift vp to heaven, and with sighs and groanings, as men that were grieved for their sinnes and offences. Although this contrition was onelie for feare of the corporal punishment that was given them, and not for any feare of the eternall, for they certainely beleeved there was no such severe punishment in the other life.

And, therefore, they offered themselves voluntarily to death, holding opinion that it is to all men an assured rest. The first day of the feast of this idoll Tezcatlipuca being come, all they of the Citty assembled together in a court to celebrate likewise the feast of the Kalender, whereof wee have already spoken, which was called Toxcoalt, which signifies a drie thing; which feast was not made to any other end, but to demaund rain, in the same manner that we solemnise the Rogations; and this feast was alwayes in Maie, which is the time that they have most neede of raine in those countries. They beganne to celebrate it the ninth of Maie, ending the nineteenth. The last day of the feast the Priestes drew foorth a litter well furnished with curtins and pendants of diverse fashions. This litter had so many armes to holde by as there were ministers to carry it. All which came foorth besmeered with black and long haire, halfe in tresses with white strings, and attyred in the livery of the idoll. Upon this litter they set the personage of the idoll appoynted for this feast, which they called the resemblance of their God Tezcalipuca, and taking it upon their shoulders they broght it openly to the foote of the stairs; then came forth the yong men and maidens of the Temple, carrying a great cord wreathed of chaines of roasted mays, with the which they invironed the Litter, putting a chaine of the same about the idolles necke, and a garland vppon his head. They called the cord Toxcalt, signifying the drought and barrennesse of the time. The yoong men came wrapped in redde curtines, with garlands and chains of roasted mays. The maides were clothed in new garments, wearing chaines about their neckes of roasted mays; and vpon their heads myters made of rods covered with this mays, they had their feete covered with feathers, and their armes and cheekes painted. They brought much of this roasted mays, and the chiefe men put it vpon their heads, and about their neckes, taking flowers in their handes. The idoll being placed in his litter, they strewed round about a great quantitie of the boughes of Manguey, the leaves whereof are large and pricking.

This litter being set vpon the religious mens shoulders, they carryed it in procession within the circuite of the Court, two Priests marching before with censors, giving often incense to the idol, and every time they gave incense they lifted vp their armes as high as they could to the idoll, and to the Sunne, saying, that they lifted vp their praiers to heaven, even as the smoke ascended on high. Then all the people in the Court turned round to the place whither the idoll went, every one carrying in his hand new cords of the threed of manguey, a fadome long, with a knotte at the end, and with them they whipped themselves vppon the shoulders; even as they doe heere vppon holy Thurseday. All the walles of the Court and battlements were full of boughs and flowers, so fresh and pleasaunt, as it did give a great contentment. This procession being ended, they brought the idoll to his vsual place of abode, then came a great multitude of people with flowres, dressed in diverse sortes, wherewith they filled the temple and all the court, so as it seemed the ornament of an Oratory. All this was putte in order by the priests, the yoong men administring these things vnto them from without. The chappell or chamber where the idoll remayned was all this day open without any vaile.

This done, every one came and offered curtines, and pendants of sendal, precious stones, iewells, insence, sweete wood, grapes, or eares of Mays, quailes: and, finally, all they were accustomed to offer in such solemnities. Whenas they offered quailes, (which was the poore mans offering,) they used this ceremonie, they delivered them to the priestes, who taking them, pulled off their heads, and caste them at the foote of the Altare, where they lost their bloud, and so they did of all other things which were offered. Every one did offer meate and fruite according to their power, the which was laid at the foote of the altar, and the Ministers gathered them vp, and carried them to their chambers. This solemne offering done, the people went to dinner, every one to his village or house, leaving the feast suspended vntil after dinner. In the meanetime, the yong men and maidens of the temple, with their ornaments, were busied to serve the idoll, with all that was appointed for him to eate: which meate was prepared by other women, who had made a vow that day to serve the idoll. And, therefore, such as had made this vow, came by the point of day, offering themselves to the Deputies of the Temple, that they might command them what they would have done, the which they did carefully performe. They did prepare such varietie of meates, as it was admirable. This meate being ready, and the hour of dinner come, all these virgins went out of the Temple in procession, every one carrying a little basket of bread in her hand, and in the other, a dish of these meates; before them marched an old man, like to a steward, with a pleasant habite, he was clothed in a white surples downe to the calves of his legges; vpon a doublet without sleeves of red leather, like to a iacket, he carried wings insteede of sleeves, from the which hung broade ribbands, at the which did hang a small calibash or pumpion, which was covered with flowers, by little holes that were made in it, and within it were many superstitious things. This old man, thus attyred, marched very humbly and heavily before the preparation, with his head declining: and comming neere the foote of the staires, he made a great obeisance and reverence. Then going on the one side, the virgins drew neere with the meate, presenting it in order one after another, with great reverence. This service presented, the old man returned as before, leading the virgins into their convent. This done, the yong men and ministers of the Temple came forth and gathered vp this meate, the which they carried to the chambers of the chiefe Priests of the Temple, who had fasted five daies, eating onely once a day, and they had also abstained from their wives, not once going out of the Temple in these five daies. During the which, they did whippe themselves rigorously with cordes, they did eate of this divine meate (for so they called it), what they could, neither was it lawfull for any other to eate thereof. All the people having dined, they assembled againe in the court to see the ende of the feast, whither they brought a captive, which by the space of a whole yeàre, had represented the idoll, being attyred, decked, and honoured as the idoll it selfe, and doing all reverence vnto him, they delivered him into the handes of the sacrificers, who at that instant presented themselves, taking him by the feete and handes. The Pope did open his stomacke, and pull out his hart, then did he lift vp his hands as high as he could, shewing it to the Sunne, and to the idoll, as hath beene said. Having thus sacrificed him that represented the idoll, they went into a holy place appointed for this purpose, whither came the yong men and virgins of the Temple with their ornaments, the which being put in order, they danced and sung with drummes and other instruments, on the which the chiefe Priests did play and sound. Then came all the Noblemen with ensignes and ornaments like to the yong men, who danced round about them. They did not usually kill any other men that day, but him that was sacrificed, yet every fourth yeare they had others with him, which was in the yeare of Iubile and full pardons. After Sun set, every one being satisfied with sounding, eating, and drinking, the virgins went al to their convent, they took great dishes of earth full of bread mixt with hony, covered with small panniers, wrought and fashioned with dead mens heads and bones, and they carried the collation to the idoll, mounting vp to the court, which was before the doore of the Oratorie: and having set them downe, they retired in the same order as they came, the steward going still before. Presently came forth all the yong men in order, with canes or reedes in their handes, who beganne to runne as fast as they could to the toppe of the staires of the Temple, who should come first to the dishes of the collation. The Elders or chiefe Priests observed him that came first, second, third, and fourth, without regarding the rest. This collation was likewise all carried away by the yong men as great relicks. This done, the foure that arrived first were placed in the midst of the Antients of the Temple, bringing them to their chambers with much honour, praising them, and giving them ornaments; and from thence forth they were respected and reverenced as men of marke. The taking of this collation being ended, and the feast celebrated with much ioy and noise, they dismissed all the yong men and maides which had served the idoll: by meanes whereof they went one after another, as they came forth. All the small children of the colledges and schooles were at the gate of the court, with bottomes of rushes and hearbes in their hands, which they cast at them, mocking and laughing, as of them that came from the service of the idoll; they had libertie then to dispose of themselves at their pleasure, and thus the feast ended.

CHAP. XXX.—*Of the Feast of Marchants, which those of Cholutecas did celebrate.*

Although I have spoken sufficiently of the service the Mexicaines did vnto their gods, yet will I speak something of the feast they called Quetzacoaatl, which was the god of riches, the which was solemnised in this maner. Fortie daies before the Marchants bought a slave well proportioned, without any fault or blemish, either of sickenes or of hurte, whom they did attyre with the ornaments of the idoll, that he might represent it fortie daies. Before his clothing they did clense him, washing him twice in a lake, which they called the lake of the gods; and being purified, they attyred him like the idoll. During these forty daies, hee was much respected for his sake whom he represented. By night they did imprison him (as hath beene said) lest he should flie, and in the morning they took him out of prison, setting him vpon an eminent place, where they served him, giving him exquisite meates to eate. After he had eaten, they put a chaine of flowers about his necke, and many nosegaies in his hands. Hee had a well appointed guard, with much people to accompany him. When he went through the Cittie, he went dancing and singing through all the streetes, that hee might bee knowne for the resemblance of their god, and when hee beganne to sing, the women and little children came forth of their houses to salute him, and to offer vnto him as to their god. Two old men of the Antients of the Temple came vnto him nine daies before the feast, and humbling themselves before him, they said with a low and submisse voyce, Sir, you must vnderstand that nine daies hence the exercise of dancing and singing doth end, and thou must then die; and then he must answer in a good houre.[1] They call this ceremony Neyòlo Maxilt Ileztli, which is to say, the advertisement;[2] and when they did thus advertise him, they took very carefull heede whether hee were sad, or if he danced as ioyfully as he was accustomed, the which if he did not as cheerefully as they desired, they made a foolish superstition in this maner. They presently tooke the sacrificing rasors, the which they washed and clensed from the blood of men which remained of the former sacrifices. Of this washing they made a drinke mingled with another liquor made of Cacao, giving it him to drinke; they said that this would make him forget what had been said vnto him, and would make him in a maner incensible, returning to his former dancing and mirth. They said, moreover, that he would offer himself cheerfully to death, being inchanted with this drinke. The cause why they sought to take from him this heavinesse, was, for that they held it for an ill augure, and a fore-telling of some

[1] " Y el avia de responder que fuesse mucho de norabuena."
[2] " El apercebimiento."

great harme. The day of the feast being come, after they had done him much honor, sung, and given him incense, the sacrificers took him about midnight and did sacrifice him, as hath been said, offering his heart vnto the Moone, the which they did afterwardes cast against the idoll, letting the bodie fall to the bottome of the staires of the Temple, where such as had offered him took him vp, which were the Marchants, whose feast it was. Then having carried him into the chiefest mans house amongst them, the body was drest with divers sawces, to celebrate (at the breake of day) the banquet and dinner of the feast, having first bid the idoll good morrow, with a small dance, which they made whilst the day did breake, and that they prepared the sacrifice. Then did all the Marchants assemble at this banket, especially those which made it a trafficke to buy and sell slaves, who were bound every yeare to offer one, for the resemblance of their god. This idoll was one of the most honoured in all the land; and therefore the Temple where he was, was of great authoritie. There were threescore staires to ascend vp vnto it, and on the toppe was a court of an indifferent largenesse, very finely drest and plastered, in the midst whereof was a great round thing like vnto an Oven, having the entrie low and narrow, so as they must stoope very low that should enter into it. This Temple had chambers and chappels as the rest, where there were convents of Priests, yong men, maides, and children, as hath been said; and there was one Priest alone resident continually, the which they changed weekely. For although there were in every one of these temples three or foure Curates or Ancients,[1] yet did every one serve his weeke without parting. His charge that weeke (after he had instructed the children) was to strike vp a drumme every day at the Sunne setting, to the same end that we are accustomed to ring to evensong. This drumme was such as they might heare the sound thereof through-

---

[1] "Curas o Dignidades."

out all the partes of the Cittie, then every man shut vp his merchandise, and retired vnto his house, and there was so great a silence, as there seemed to be no living creature in the Towne. In the morning whenas the day beganne to breake, they beganne to sound the drumme, which was a signe of the day beginning, so as travellers and strangers attended this signall to beginne their iournies, for till that time it was not lawfull to goe out of the cittie.

There was in this temple a court of a reasonable greatnes, in the which they made great dances and pastimes, with games or comedies the day of the idolls feast; for which purpose there was in the middest of this court a theatre of thirty foote square, very finely decked and trimmed, the which they decked with flowers that day, with all the arte and invention that mought be, beeing invironed round with arches of divers flowers and feathers, and in some places there were tied many small birds, connies, and other tame beasts. After dinner all the people assembled in this place, and the players presented themselves, and played comedies: some counterfeit the deafe and the rheumatike, others the lame, some the blinde, and without handes, which came to seeke for cure of the idoll: the deafe answered confusedly, the rheumatike did cough, the lame halted, telling their miseries and griefes, wherewith they made the people to laugh; others came foorth in the forme of little beasts, some were attired like snailes, others like toades, and some like lizardes: then meeting together, they tolde their offices, and every one retyring to his place, they sounded on small flutes, which was pleasant to heare. They likewise counterfeited butterflies and small birdes of diverse colours, and the children of the Temple represented these formes; then they went into a little forrest planted there for the nonce, where the Priests of the Temple drew them foorth with instruments of musicke. In the meane time they vsed many pleasant speeches, some in propounding, others in defending, wherewith the assistants were pleasantly intertained. This doone, they made a maske or mummerie with all these personages, and so the feast ended: the which were vsually doone in their principall feasts.

CHAP. XXXI.—*What profit may be drawne out of this discourse of the Indians superstitions.*

This may suffice to vnderstand the care and paine the Indians tooke to serve and honour their Idolls, or rather the divell: for it were an infinite matter, and of small profit, to report every thing that hath passed, for that it may seeme to some needlesse to have spoken thus much: and that it is a losse of time, as in reading the fables that are fained by the Romaines of Knighthoode. But if such as holde this opinion will looke wel into it, they shall finde great difference betwixt the one and the other: and that it may be profitable, for many considerations, to have the knowledge of the customs and ceremonies the Indians vsed: first, this knowledge is not only profitable, but also necessary in those countries where these superstitions have been practised, to the end that Christians, and the maisters of the Law of Christ, may knowe the errours and superstitions of the Antients, and observe if the Indians vse them not at this day, either secretely or openly. For this cause many learned and worthy men have written large Discourses of what they have found: yea, the Provinciall counsells have commaunded them to write and print them, as they have doone in Lima, where hath beene made a more ample Discourse than this. And therefore it importeth for the good of the Iudians, that the Spaniardes being in those parts of the Indies, should have the knowledge of all these things. This Discourse may likewise serve the Spaniards there, and all others whersoever, to give infinite thankes to God our Lord, who hath imparted so great a benefite vnto vs, giving them his holy Lawe, which is most iust, pure, and altogether profitable. The which we may well know, comparing it with the lawes of Sathan, where so many wretched people have lived so miserably. It may likewise serve to discover the pride, envy, deceipts, and ambushes of the Divell, which he practiseth against those hee holdes captives, seeing on the one side hee seekes to imitate God, and make comparison with him and his holy Lawe; and on the other side, hee dooth mingle with his actions so many vanities, filthinesse, and cruelties, as hee that hath no other practise but to sophisticate and corrupt all that is good. Finally, hee that shall see the darkenes and blindenes wherein so many Provinces and Kingdoms have lived so long time, yea and wherein many Nations, and a great part of the world live yet, deceived with the like trumperies, he can not (if he have a Christians heart) but give thankes to the high God, for such as hee hath called out of so great darkenes, to the admirable light of his Gospel: beseeching the vnspeakeable charitie of the Creator to preserve and increase them in his knowledge and obedience, and likewise be grieved for those that follow still the way of perdition. And that in the end hee beseech the Father of Pitty to open vnto them the treasures and riches of Iesus Christ, who with the Father and Holy Ghost raignes in all Ages. Amen.

THE SIXT BOOKE

Of the Naturall and Morall Historie of the Indies.

CHAP. I.—*That they erre in their opinion, which holde the Indians to want iudgement.*

LIB. VI.

HAVING treated before of the religion the Indians vsed, I pretend to discourse in this Booke of their customs, policy, and government, for two considerations: the one is to confute that false opinion many doe commonly holde of them, that they are a grose and brutish people, or that they have so little vnderstanding, as they scarce deserve the name of anie. So as many excesses and outrages are committed vpon them, vsing them like bruite beasts, and reputing them vnworthy of any respect; which is so common and so dangerous an errour (as they know well who with any zeale and consideration have travelled amongst them, and that have seene and observed their secrets and counsells). And moreover, for the small regard many make of these Indians, who presume to knowe much, and yet are commonly the most ignorant and presumptuous. I finde no better meanes to confound this pernicious opinion, then in relating their order and maner, whenas they lived vnder their owne lawes, in which, although they had many barbarous things, and without ground, yet had they many others worthy of great admiration, whereby wee may vnderstand, that they were by nature capable to receive any good instructions: and besides, they did in some things passe many of our commonweales. It is no matter of marvell if there were so great and grose faults amongst them, seeing there hath been likewise amongst the most famous Law-givers and Philosophers (yea, without exception, Lycurgus and Plato), and amongest the wisest common-wealths, as the Romanes and Athenians, where wee may finde things so full of ignorance, and so worthy of laughter, as in trueth if the commonweales of the Mexicaines, or of the Yncas, hadde beene knowne in those times of the Romans and the Greekes, their lawes and governments had been much esteemed by them. But we at this day little regarding this, enter by the sword, without hearing or vnderstanding; perswading our selves that the Indians affaires deserve no other respect, but as of venison that is taken in the forrest, and broght for our vse and delight.

The most grave and diligent, which have searched and attained to the knowledge of their secrets, customs, and antient government, holde another opinion, and admire the order and discourse that hath been betwixt them. Of which number is Polo Ondegardo, whome I vsually followe in the discourse of matters of Peru, and for these of Mexico Juan de Tobar, who had a Prebend in the Church of Mexico, and is now of our company of Iesuites, who by the commaundement of the viceroy Don Martin Henriques,[1] have made a diligent and a large collection of the histories of that nation, and many other grave and notable personages, who, both by word and writing, have sufficiently informed me of all those things I shall here set downe. The other end, and the good which may followe by the knowledge of the lawes, customes, and government of the Indians, is, that wee may helpe and governe them with the same lawes and customes,

[1] Second son of Don Francisco Henriquez y Almansa, first Marquis of Alcanizes, by Doña Isabel de Ulloa. He was Viceroy of Mexico from 1568 to 1580, and of Peru from 1581 to 1583. He died at Lima on March 12th, 1583.

for that they ought to be ruled according to their owne lawes and priviledges, so farre foorth as they doe not contradict the Lawe of Christ, and his holy Church, which ought to be maintained and kept as their fundamentall lawes. For the ignorance of laws and customes hath bred many errours of great importance, for that the Governours and Iudges knowe not well how to give sentence, nor rule their subjects. And besides, the wrong which is doone vnto them against reason, it is preiudiciall and hurtefull vnto our selves; for thereby they take occasion to abhorre vs, as men both in good and in evill alwayes contrary vnto them.

CHAP. II.—*Of the method of computing time, and the Kalendar the Mexicaines vsed.*

And to beginne then by the division and supputation of times which the Indians made, wherein truely wee may well perceive the great signes of their vivacitie and good vnderstanding. I will first shew in what sorte the Mexicaines counted and divided their yeere, their moneths, their kalender, their computations, their worldes and ages. They divided the yeare into eighteene moneths, to which they gave twentie dayes, wherein the three hundred and three score days are accomplished, not comprehending in any of these moneths the five dayes that remaine, and make the yeare perfect. But they did reckon them aparte, and called them the dayes of nothing: during the which, the people did not any thing, neither went they to their Temples, but occupied themselves only in visiting one another, and so spent the time: the sacrificers of the Temple did likewise cease their sacrifices. These five dayes being past, they beganne the computation of the yeare, whereof the first moueth and the beginning was in March, when the leaves beganne to growe greene, although they tooke three dayes of the moneth of February; for the first day of their yeere was, as it were, the sixe and twentie day of February, as appeareth by their kalender, within the which ours is likewise comprehended and contained with a very ingenious Arte, which was made by the antient Indians that knew the first Spaniardes. I have seene this Kalender, and have it yet in my custody, which well deserveth the sight, to vnderstand the discourse and industry the Mexicaine Indians had. Every one of these eighteene monethes had his proper name, and his proper picture, the which was commonly taken of the principall feast that was made in that moneth, or from the diversitie of tymes, which the yeere caused in that moneth. They had in this Kalender certaine dayes marked and distinguished for their feasts. And they accompted their weekes by thirteene dayes, marking the dayes with a Zero or cipher, which they multiplied vnto thirteene, and then beganne to count, one, two, etc. They did likewise marke the yeares of these wheeles with foure signes or figures, attributing to every yeare a peculiar signe, wherof one was of a house, an other of a conny, the third of a reede, and the fourth of a flint. They painted them in this sort, noting by those figures the yeare that did runne, saying of so many houses, of so many flints of such a wheele, happened such a thing. For we must vnderstand that their wheele, which was an age, contained foure weekes of yeares, every weeke containing thirteene yeares, which in all made fiftie twoo yeares. In the midst of this wheele they painted a Sunne, from the which went foure beames or lines in crosse to the circumference of the wheele; and they made their course, even as the circumference was divided into foure equall partes, every one with his line, having a distinct colour from the rest, and the foure colors were greene, blew, red, and yellow: every portion of these foure had thirteene separations which had all their signes or particular figures, of a house, a conny, a reed, or a flint, noting by every signe a yeare, and vppon the head of this signe they painted what had happened that yeare.

And therefore I did see in the Kalender mentioned the yeare when the Spaniards entered Mexico, marked by the picture of a man clad in red, after our manner, for such was the habite of the first Spaniard, whome Fernand Cortes sent at the end of the two and fifty years, which finished the wheele. They vsed a pleasant ceremony, which was the last night they didde breake all their vesselles and stuffe, and put out their fire, and all the lights, saying, that the worlde should end at the finishing of one of these wheeles, and it might be at that time : for (said they), seeing the worlde must then end, what neede is there to provide meate to eate, and therefore they had no further neede of vessel nor fire. Vpon this conceit they passed the night in great feare, saying it might happen there would be no more day, and they watched very carefully for the day; but when they saw the day beginne to breake, they presently beat manie drummes, and sounded cornets, flutes, and other instruments of ioy and gladnesse, saying, that God did yet prolong the time with another age, which were fiftie two yeares. And then beganne an other wheele. The first day and beginning of this age they took new fire, and bought new vesselles to dresse their meate, and all went to the high Priest for this new fire, having first made a solemne sacrifice, and given thanks for the comming of the day, and prolongation of an other age. This was their manner of accounting their yeares, moneths, weekes, and ages.

CHAP. III.—*How the Kings Yncas accounted the yeares and moneths.*

Although the computation of time practised amongst the Mexicaines bee ingenious enough and certaine, for men that had no learning; yet, in my opinion, they wanted discourse and consideration, having not grounded their computation according vnto the course of the moone, nor distributed their months accordingly, wherein those of Peru have far surpassed them: for they divided their yeare into as many dayes, perfectly accomplished as we do heere, and into twelve moneths or moones, in the which they imployed and consumed the eleven daies that remaind of the moone, as Polo writes. To make the computation of their yeare sure and certaine, they vsed this industry; vppon the mountaines which are about the citty of Cuzco (where the Kings Yncas held their court, beeing the greatest sanctuary of those realmes, and as we should say an other Rome), there were twelve pillars set in order, and in such distaunce the one from the other, as every month one of these pillars did note the rising and setting of the sunne. They called them *Succanga*,[1] by meanes whereof they taught and shewed the feasts, and the seasons fitte to sowe and reape, and to do other things. They did certaine sacrifices to these pillars of the sunne. Every month had his proper name and peculiar feasts. They beganne the yeare by Ianuary, as wee doe. But since, a king Ynca called Pachacutec,[2] which signifies a reformer of time, beganne their yeare by December, by reason (as I coniecture) that then the Sunne returneth

[1] *Sucanca. Suca* is a ridge or furrow in Quichua. *Sucani*, "I make furrows". *Sucanca* is the future passive participle, "that which is about to be furrowed"; possibly referring to the alternate light and shadow caused by the sunlight between the pillars; making the ground appear in ridges.—See *G. de la Vega*, i, p. 178.

[2] *Pacha*, time; *Cutini*, I overturn, or reform.

LIB. VI. from the last poynt of Capricorne, which is the tropike neerest vnto them. I know not whether the one or the other have observed any Bisexte, although some holde the contrary. The weekes which the Mexicaines did reckon were not properly weekes, being not of seaven dayes: the Yncas likewise made no mention thereof, which is no wonder, seeing the account of the weeke is not grounded vpon the course of the sunne, as that of the yeare, nor of the moone, as that of the month; but among the Hebrewes it is grounded vpon the creation of the world, as Moses reporteth; and amongest the Greekes and Latins vpon the number of the seven planets, of whose names the dayes of the weeke have taken their denomination; yet was it much for those Indians, being men without bookes and learning, to have a yeare, seasons, and feasts, so well appoynted as I have sayd.

CHAP. IV.—*That no nation of the Indies hath beene found to have had the vse of letters.*

Letters were invented to signifie properly the words we do pronounce, even as woordes (according to the Philosopher) are the signes and demonstrations of mans thoughtes and conceptions. And both the one and the other (I say the letters and words) were ordained to make things knowne. The voyce for such as are present, and letters for the absent, and such as are to come. Signes and markes which are not properly to signifie wordes but things, cannot be called, neyther in trueth are they letters, although they be written, for wee can not say that the Picture of the sunne be a writing of the sunne, but onely a picture; the like may be saide of other signes and characters, which have no resemblance to the thing, but serve onely for memorie: for he that invented them did not ordaine them to signifie wordes, but onely to note the thing: neyther do they call those characters, letters, or writings, as indeede they are not, but rather ciphers or remembraunces, as those be which the Spherists or Astronomers do vse, to signifie divers signes or planets of Mars, Venus, Iupiter, etc.

Such characters are ciphers, and no letters: for what name soever Mars may have in Italian, French, or Spanish, this character doth alwaies signifie it, the which is not found in letters: for, althogh they signify the thing, yet is it by meanes of words. So, as they which know not the tongue, vnderstand them not: as, for example, the Greekes nor the Hebrews, cannot conceive what this word Sol doth signifie, although they see it written; for that they vnderstand not the Latine word: so as writing and letters are onely practised by them, which signifie words therewith. For if they signifie things mediately, they are no more letters nor writings, but ciphers and pictures: whereby we may observe two notable things. The one, that the memory of histories and antiquities may bee preserved by one of these three meanes, either by letters and writings, as hath beene vsed amongst the Latines, Greekes, Hebrews, and many other Nations; or by painting, as hath beene vsed almost throughout all the world, for it is said in the second Nicene Counsell, "Painting is a booke for fooles which cannot reade": or by ciphers and characters, as the cipher signifies the number of a hundred, a thousand, and others, without noting the word of a hundred or a thousand. The other thing we may observe thereby is that which is propounded in this chapter, which is, that no Nation of the Indies discovered in our time, hath had the vse of letters and writings, but of the other two sortes, images and figures. The which I observe, not onely of the Indies of Peru and New Spaine, but also of Iappon and China. And although this may seeme false to some, seeing it is testified by the discourses that have beene written, that there are so great

Libraries and Vniversities in China and Iappon, and that mention is made of their Chapas, letters, and expeditions, yet that which I say is true, as you may vnderstand by the discourse following.

CHAP. V.—*Of the fashion of Letters and Bookes the Chinois vsed.*

There are many which thinke, and it is the most common opinion, that the writings which the Chinois vsed are letters, as those we vse in Europe, and that by them wee may write wordes and discourses, and that they only differ from our letters and writings in the diversitie of characters, as the Greekes differ from the Latines, and the Hebrews from the Chaldees. But it is not so, for they have no Alphabet, neither write they any letters, but all their writing is nothing else but painting and ciphering: and their letters signifie no partes of distinctions as ours do, but are figures and representations of things, as of the Sunne, of fire, of a man, of the sea, and of other things. The which appears plainely, for that their writings and *chapas* are vnderstood of them all, although the languages the Chinois speake are many and very different, in like sort as our numbers of ciphers are equally vnderstoode in the Spanish, French, and Arabian tongues: for this figure 8, wheresoever it be, signifies eight, although the French call this number of one sort and the Spaniards of another. So as things being of themselves innumerable, the letters likewise or figures which the Chinois vse to signifie them by, are in a maner infinite: so as he that shall reade or write at China (as the Mandarins doe) must know and keepe in memory at the least fourescore and five thousand characters or letters, and those which are perfect herein know above sixscore thousand. A strange and prodigious thing; yea, incredible, if

it were not testified by men worthy of credite, as the fathers
of our company who are there continually, learning their
language and writing, wherein they have studied day and
night above tenne yeares, with a continuall labour for the
charitie of Christ and the desire of salvation of soules, prevailed in them above all this labour and difficultie. For
this reason, learned men are so much esteemed in China, for
the difficultie there is to conceive them: and those only
have the offices of Mandarins, Governours, Iudges, and
Captaines. For this cause the fathers take great pains to
instruct their children to reade and write. There are many
of these schooles where the children are taught, where the
masters teach them by day, and the fathers at home by
night: so as they hurt their eyes much, and they whippe
them often with reedes, although not so severely as they doe
offenders. They call it the Mandarin tongue, which requires
a mans age to be conceived. And you must vnderstand
that, although the tongue which the Mandarins speake bee
peculiar and different from the Vulgar, which are many,
and that they studie it, as they doe Latine and Greeke
heere, and that the learned only throghout all China do
vnderstand it: so it is notwithstanding that all that is
written in it, is vnderstood in all tongues: and although all
the Provinces doe not vnderstand one another by speaking,
yet by writing they doe: for there is but one sort of figures
and characters for them all, which signifie one thing, but
not the same word and prolation: seeing (as I have said)
they are onely to denote the things and not the worde, as
we may easily vnderstand by the examples of numbers in
ciphering. And they of Iappon and the Chinois do reade
and vnderstand well the writings one of another, although
they be divers Nations and different in tongue and language. If they speake what they reade or write, they
should not bee vnderstood. Such are the letters and bookes
the Chinois vse, being so famous in the world. To make

their impressions, they grave a boord or plank with the figures they will print, then do they stampe as many leaves of paper as they lift, of the same sort as they have made their pictures, the which are graven in copper or wood. But a man of iudgement may aske, how they could signifie their conceptions by figures, which approached neere or resemble the thing they would represent? As to say, the Sunne heats, or that he hath beheld the Sunne, or the day is of the Sunne. Finally, how it were possible to denote by the same figures the case, the coniunction, and the articles, which are in many tongues and writings? I answer therevnto, that they distinguish and signifie this varietie by certaine points, strikes, and dispositions of the figure. But it is difficult to vnderstand how they can write proper names in their tongue, especially of strangers, being things they have never seene, and not able to invent figures proper vnto them. I have made triall thereof, being in Mexico with certain Chinois, willing them to write this proposition in their language, "Ioseph de Acosta has come from Peru", and such like: wherevpon the Chinois was long pensive, but in the end hee did write it, the which other Chinois did after reade, although they did vary a little in the pronountiation of the proper name. For they vse this devise to write a proper name: they seeke out something in their tongue that hath resemblance to that name, and set downe the figure of this thing. And as it is difficult among so many proper names to finde things to resemble them in the prolation, so is it very difficult and troublesome to write such names. Vpon this purpose, father Alonso Sanchez told vs that when he was in China, being led into divers Tribunall seates, from Manderin to Manderin, they were long in putting his name in writing in their *chapas*, yet in the end they did write it after their maner, and so ridiculously, that they scarce came neere to the name: and this is the fashion of letters and writings which the Chinois vsed. That of

CHINESE LEARNING. 401

the Iapponois approched very neere, although they affirme that the Noblemen of Iappon that came into Europe did write all things very easily in their language were they of our proper names: yea, I have had some of their writings shewed me, whereby it seemes they should have some kinde of letters, although the greatest part of their writings be by the characters and figures, as hath bin saide of the Chinois.

CHAP. VI.—*Of the Schooles and Vniversities of China.*

The fathers of our Company say that they have not seene in China any great schooles or vniversities of Philosophie, and other naturall sciences, beleeving there is not any, but that all their studie is in the Mandarin tongue, which is very ample and hard, as I have said; and what they studie bee things written in their owne tongue, which be histories of sects, and opinions, of civill lawes, of morall proverbs, of fables, and many other such compositions that depend thereon. Of divine sciences they have no knowledge, neither of naturall things, but some small remainders of straied propositions, without art or methode, according to everie mans witte and studie. As for the Mathematikes, they have experience of the celestiall motions, and of the starres. And for Phisicke, they have knowledge of herbs, by means wherof they cure many diseases, and vse it much. They write with pencils, and have many books written with the hand, and others printed, but in a bad order. They are great plaiers of comedies, the which they perform with great preparation of theaters, apparel, bels, drums, and voices. Some fathers report to have seen comedies which lasted ten or twelve dayes and nights, without any want of comedians, nor company to beholde them. They doe make many different sceanes, and whilst some act the others feede and sleep. In these comedies they do commonly treate of morall things, and of good examples, intermingled with pleasant devices. This is the summe of that which our men report of the letters and exercises of them of China, wherein wee must confesse to be much wit and industrie. But all this is of small substance, for in effect all the knowledge of the Chinois tendes onely to read and write, and no farther, for they attaine to no high knowledge. And their writing and reading is not properly reading and writing, seeing their letters are no letters that can represent wordes, but figures of innumerable things, the which cannot be learned but in a long time, and with infinite labour. But in the end, with all their knowledge, an Indian of Peru or Mexico that hath learned to read and write knowes more than the wisest Mandarin that is amongst them: for that the Indian with foure and twentie letters which he hath learned will write all the wordes in the world, and a Mandarin with his hundred thousand letters will be troubled to write some proper name, as of Martin, or Alonso, and with greater reason he shall be lesse able to write the names of things he knowes not. So as the writing in China is no other thing but a maner of painting or ciphring.

CHAP. VII.—*Of the fashion of letters and writings which the Mexicaines vsed.*

We finde amongest the Nations of New Spaine a great knowledge and memorie of antiquitie, and therefore, searching by what meanes the Indians had preserved their Histories and so many particularities, I learned that although they were not so subtill and curious as the Chinois and those of Iappon, yet had they some kinde of letters and bookes amongest them whereby they preserved (after their manner) the deeds of their predecessors. In the province of Yu-catan, where the Bishopricke is, which they call of

Honduras, there were bookes of the leaves of trees, folded and squared after their manner, in the which the wise Indians contained the distribution of their times, the knowledge of the planets, of beasts and other naturall things, with their antiquities, a thing full of great curiositie and diligence. It seemed to some Pedant that all this was an inchantmênt and magicke arte, who did obstinately maintaine that they ought to be burnt, so as they were committed to the fire. Which since, not onely the Indians found to be ill done, but also the curious Spaniards, who desired to know the secrets of the countrey. The like hath happened in other things, for our men thinking that all was but superstition have lost many memorialls of ancient and holy things, which might have profited·much. This proceedeth of a foolish and ignorant zeale, who not knowing, nor seeking to knowe what concerned the Indians, say preiudicately that they are all but witchcrafts, and that all the Indians are but drunkards, incapable to know or learne anything. For such as would be curiously informed of them have found many things worthy of consideration. One of our company of Iesuites, a man very witty and wel experienced, did assemble in the province of Mexico the Antients of Tescuco, of Talla, and of Mexico, conferring at large with them, who shewed unto him their books, histories and kalenders, things very woorthy the sight, bicause they had their figures and hierogliphicks, whereby they represented things in this manner: Such as had forme or figure were represented by their proper images, and such as had not any were represented by characters that signified them, and by this meanes they figured and writ what they would. And to observe the time when anything did happen they had those painted wheeles, for every one of them contained an age, which was two and fifty years, as hath beene said; and of the side of those wheeles they did paint with figures and characters, right against the yeare, the memorable things that happened therein. As they noted the yeare whenas the Spaniards entred their Countrey, they painted a man with a hatte and a red ierkin vpon the signe of the reede, which did rule then, and so of other accidents. But for that their writings and characters were not sufficient, as our letters and writings be, they could not so plainly expresse the words, but onely the substance of their conceptions. And forasmuch as they were accustomed to reherse Discourses and Dialogues by heart, compounded by their Oratours and auntient Rhethoritians, and many Chapas made by their Poets (which were impossible to learne by their Hierogliphickes and Characters), the Mexicaines were very curious to have their children learne those dialogues and compositions by heart. For the which cause they had Schooles, and as it were Colledges or Seminaries, where the Auncients taught children these Orations, and many other things, which they preserved amongst them by tradition from one to another as perfectly as if they had beene written; especially the most famous Nations had a care to have their children (which had any inclination to be Rhetoritians, and to practise the office of Orators) to learne these Orations by heart: So as when the Spaniardes came into their Countrey, and had taught them read and write our letters, many of the Indians then wrote these Orations, as some grave men doe witnes that had read them. Which I say, for that some which shall haply reade these long and eloquent discourses in the Mexicaine Historie will easilie beleeve they have beene invented by the Spaniardes, and not really taken and reported from the Indians. But having knowne the certaine trueth, they will give credite (as reason is) to their Histories. They did also write these Discourses after their manner, by Characters and Images: and I have seene, for my better satisfaction, the *Pater noster*, *Ave Maria*, and *Simboll*, and the generall confession of our faith, written in this manner by the Indians.

And in trueth, whosoever shall see them will wonder thereat. For to signifie these words, I, a sinner, do confesse my self, they painted an Indian vpon his knees at a religious mans feete, as one that confesseth himselfe: and for this, to God most mighty, they painted three faces, with their crownes, like to the Trinitie; and to the glorious Virgine Marie, they painted the face of our Lady, and halfe the body of a little childe; and for S. Peter and S. Paul, heads with crowns, and a key with a sword; and whereas images failed, they did set characters, as "Wherein I have sinned, etc.", whereby wee may conceive the quickenesse of spirite of these Indians, seeing this manner of writing of our prayers and matters of faith hath not been taught them by the Spaniards, neither could they have done it if they had not had an excellent conception of that was taught them. And I have seene in Peru a confession of sinnes, brought by an Indian, written in the same sorte, with pictures and characters, painting every one of the tenne Commandments after a certaine manner where there were certaine markes like ciphers, which were the sinnes he had committed against the Commandments. I nothing doubt but if any of the most sufficient Spaniards were imployed to make memorialles of the like things by their images and markes, they would not attaine vnto it in a whole year, no not in tenne.

Chap. viii.—*Of Registers and the manner of reckoning which the Indians of Peru vsed.*

Before the Spaniards came to the Indies, they of Peru had no kinde of writing, either letters, characters, ciphers, or figures, like to those of China and Mexico: yet preserved they the memory of their Antiquities, and maintained an order in all their affairs of peace, warre, and pollicie, for that they were carefull observers of traditions from one to another, and the young ones learned, and carefully kept, as a holy thing, what their superiors had tolde them, and taught it with the like care to their posteritie. Besides this diligence, they supplied the want of letters and writings, partely by painting, as those of Mexico (although they of Peru were very grosse and blockish[1]), and partely, and most commonly by Quippos.[2] These Quippos are memorialls or or registers, made of bowes,[3] in the which there are diverse knottes and colours, which do signifie diverse things, and it is strange to see what they have expressed and represented by this meanes: for their Quippos serve them insteede of Bookes of histories, of lawes, ceremonies, and accounts of their affaires. There were officers appointed to keepe these Quippos, the which at this day they call Quipocamayos, the which were bound to give an account of everything, as Notaries and Registers doe heere. Therefore they fully believed them in all things, for, according to the varietie of business, as warres, pollicie, tributes, ceremonies and landes, there were sundry Quippos or braunches, in every one of the which there were so many knottes, little and great, and strings tied vnto them, some red, some greene, some blew, some white; and finally, such diversitie, that even as wee derive an infinite number of woordes from the foure and twenty letters, applying them in diverse sortes, so doe they draw innumerable woordes from their knottes and diversitie of colours. Which thing they doe in such a manner that if at this day in Peru, any Commissary come at the end of two or three years to take information vppon the life of any officer, the Indians come with their small reckonings verified, saying, that in such a village they have given him so many egges which he hath not payed for, in such a house a henne, in another two burdens of grasse for his horse, and that he hath paied but

[1] " Muy grosseras y toscas."   [2] " Quipus."
[3] " Ramales," rope's-ends.

so much mony, and remaineth debtor so much. The proofe
being presently made with these numbers of knottes and
handfulls of cords, it remaines for a certaine testimony and
register. I did see a handfull of these strings, wherein an
Indian woman carried written a generall confession of all
her life, and thereby confessed herselfe as well as I could
have done it in written paper. I asked her what those
strings meant that differed from the rest: she answered mee
they were certaine circumstaunces which the sin required to
be fully confessed. Besides these Quippos of thred, they have
an other, as it were a kinde of writing with small stones,
by means whereof they learne punctually the words they
desire to know by heart. It is a pleasant thing to see the
olde and the impotent (with a wheele made of small stones)
learne the *Pater noster,* with another the *Ave Maria,* with
another the Creede; and to remember what stone signifies
"Which was conceived by the holy-ghost", and which
" Suffered under Pontius Pilate".

It is a pleasant thing to see them correct themselves
when they doe erre; for all their correction consisteth
onely in beholding of their small stones. One of these
wheeles were sufficient to make mee forget all that I do
knowe by heart. There are a great number of these wheeles
in the Church-yards for this purpose. But it seemes a
kinde of witchcraft, to see an other kinde of Quippos, which
they make of graines of Mays, for to cast vp a hard
account, wherein a good Arithmetitian would be troubled
with his penne to make a division; to see how much every
one must contribute : they do drawe so many graines from
one side, and adde so many to another, with a thousand
other inventions. These Indians will take their graines,
and place five of one side, three of another, and eight of
another, and will change one graine of one side, and three
of another. So as they finish a certaine account, without
erring in any poynt : and they sooner submitte themselves

to reason by these Quippos, what every one ought to pay, then we can do with the penne. Hereby we may judge if they have any understanding, or be brutish: for my parte, I think they passe vs in those things wherevnto they do apply themselues.

CHAP. IX.—*Of the order the Indians holde in their writings.*

It shalbe good to adde heerevnto what we have observed touching the Indians writings; for their manner was not to write with a continued line, but from the toppe to the bottome, or in circle-wise. The Latines and Greeks do write from the left hand vnto the right, which is the vulgar and common manner we do vse. The Hebrewes contrariwise beganne at the right to the left, and therefore their bookes beganne where ours did end. The Chinois write neither like the Greeks nor like the Hebrews, but from the toppe to the bottome, for as they be no letters but whole wordes, and that every figure and character signifieth a thing, they have no neede to assemble the parts one with an other, and therefore they may well write from the toppe to the bottome. Those of Mexico for the same cause did not write in line, from one side to another, but contrarie to the Chinois, beginning below, they mounted vpward. They vsed this maner of writing, in the account of their daies, and other things which they observed. Yet when they did write in their wheels or signes, they beganne from the middest where the sun was figured, and so mounted by their yeeres vnto the round and circumference of the wheele. To conclude, wee finde four different kindes of writings, some writte from the right to the left, others from the left to the right, some from the toppe to the bottome, and others from the foote to the toppe, wherein wee may discover the diversity of mans judgment.

Chap. x.—*How the Indians dispatched their Messengers.*

To finish the maner they had of writing, some may, with reason, doubt how the Kings of Mexico and Peru had intelligence from all those realmes that were so great, or by what means they could dispatch their affaires in Court, seeing they had no vse of any letters, nor to write pacquets: wherein we may be satisfied of this doubt, when we understand that by wordes, pictures, and these memorialles, they were often advertised of that which passed. For this cause there were men of great agilitie, which served as curriers, to goe and come, whom they did nourish in this exercise of running from their youth, labouring to have them well breathed, that they might runne to the toppe of a high hill without wearines. And therefore in Mexico they gave the prize to three or foure that first mounted vp the staires of the Temple, as hath beene said in the former Booke. And in Cusco, when they made their solemne feast of Capacrayme, the novices did runne who could fastest vp the rocke of Yanacauri. And the exercise of running is generally much vsed among the Indians. Whenas there chaunced any matter of importaunce, they sent vnto the Lordes of Mexico, the thing painted, whereof they would advertise them, as they did when the first Spanish ship appeared to their sight, and when they tooke Toponchan. In Peru they were very curious of footemen, and the Ynca had them in all parts of the realme as ordinary Posts, called *Chasquis*, whereof shall be spoken in his place.

Chap. xi.—*Of the manner of governement, and of the Kings which the Indians had.*

It is apparant that the thing wherein these barbarous people shew their barbarisme, was in their governement and manner of commaund: for the more that men approch to reason, the more milde is their governement, and lesse insolent; the Kings and Lords are more tractable, agreeing better with their subiects, acknowledging them equall in nature, though inferiour in duetie and care of the commonwealth. But amongst the Barbarians all is contrary, for that their government is tyrannous, vsing their subiects like beasts, and seeking to be reverenced like gods. For this occasion many nations of the Indies have not indured any Kings or absolute and soveraigne Lords, but live in comminalities, creating and appointing Captains and Princes for certaine occasions onely, to whome they obey during the time of their charge, then after they returne to their former estates. The greatest part of this new world (where there are no settled kingdoms, nor established commonweales, neither princes nor succeeding kings) they governe themselves in this manner, although there be some Lordes and principall men raised above the common sort. In this sorte the whole Countrey of Chille is governed, where the Araucanos, those of Tucapel and others, have so many yeeres resisted the Spaniards. And in like sort all the new kingdome of Granada, that of Guatemala, the Ilandes, all Florida, Brassill, Luson, and other countries of great circuite: but that in some places, they are yet more barbarous, scarcely acknowledging any head, but all commaund and governe in common, having no other thing, but wil, violence, unreason, and disorder, so as he that most may, most commaunds. At the East Indies there are great kingdomes, well ordered and governed, as that of Siam, Bisnaga,[1] and others, which may bring to field when they please, a hundred or two hundred thousand men.

As likewise the Kingdome of China, the which in greatnes and power surpasseth all the rest, whose kings (as they report) have continued above two thousand yeares, by

"Bijaynagar."

meanes of their good order and government. But at the West Indies they have onely found two Kingdomes or setled Empires, that of the Mexicanes in New Spaine, and of the Yncas in Peru. It is not easie to be said which of the two was the mightiest Kingdome, for that Moteçuma exceeded them of Peru in buildings and in the greatnes of his court: but the Yncas did likewise exceede the Mexicaines in treasure, riches, and greatnes of Provinces. In regarde of antiquitie, the Monarchie of the Yncas hath the advantage, although it be not much, and in my opinion they have been equall in feates of armes and victories. It is most certaine that these two Kingdomes have much exceeded all the Indian Provinces discovered in this new world, as well in good order and government as in power and wealth, and much more in superstition and service of their idolls, having many things like one to an other. But in one thing they differed much, for among the Mexicaines the succession of the kingdome was by election, as the Empire of the Romans, and that of Peru was hereditarie, and they succeeded in bloud, as the Kingdomes of Fraunce and Spaine. I will therefore heereafter treate of these two governments (as the chiefe subiect and best knowne amongst the Indians) being fit for this discourse, leaving many tedious details which are not of importance.

CHAP. XII.—*Of the Government of the Kings Yncas of Peru.*

The Ynca which ruled in Peru being dead, his lawfull sonne succeeded him, and so they held him that was borne of his chiefe wife, whome they called Coya. The which they have alwaies observed since the time of an Ynca, called Yupanqui, who married his sister: for these Kings held it an honour to marry their sisters. And although they had other wives and concubines, yet the succession of the Kingdome appertained to the son of the Coya. It is true, that when the King had a legitimate brother, he succeeded before the sonne, and after him his nephew and sonne to the first. The Curacas and Noblemen held the same order of succession in their goods and offices. And after their maner they made excessive ceremonies and obsequies for the dead. They observed one custome very great and full of state, that a King which entred newly into his Kingdome should not inherite any thing of the movables, implements, and treasure of his predecessour, but hee must furnish his house new, and gather together gold, silver, and other things necessarie, not touching any thing of the deceased, the which was wholly dedicated for his Oratorie or Guaca, and for the entertainment of the family he left, the which with his of-spring was alwayes busied at the sacrifices, ceremonies, and service of the deceased King: for, being dead, they presently held him for a god, making sacrifices vnto him, images, and such like. By this meanes, there was infinite treasure in Peru: for every one of the Yncas had laboured to have his Oratorie and treasure surpasse that of his predecessors. The marke or ensigne, whereby they tooke possession of the realme, was a red rowle of wooll, more fine then silke, the which hung in the middest of his forehead: and none but the Ynca alone might weare it, for that it was as a Crowne and royall Diademe: yet they might lawfully weare a rowle hanging on the one side, neere vnto the eare, as some Noblemen did, but onely the Ynca might carry it in the middest of his forehead. At such time as they tooke this roule or wreathe, they made solemne feasts and many sacrifices, with a great quantity of vessells of gold and silver, a great number of small formes or images of sheep, made of gold and silver, great abundance of the stuffes of Cumbi,[1] well wrought, both fine and coarser, many shells of the sea of all sortes,

---

[1] "*Ccompi*", fine cloth.

many feathers, and a thousand sheepe, which must be of divers colours. Then the chiefe Priest tooke a yong child in his handes, of the age of six or eight yeares, pronouncing these wordes with the other ministers speaking to the image of Viracocha, " Lord, we offer this vnto thee, that thou maiest maintaine vs in quiet, and helpe vs in our warres, maintaine our Lord the Ynca in his greatnes and estate, that hee may alwaies increase, giving him much knowledge to governe vs." There were present at this ceremony and oath men of all partes of the Realme, and of all Guacas and Sanctuaries. And without doubt, the affection and reverence this people bare to their Kings Yncas, was very great, for it is never found that any one of his subiectes committed treason against him, for that they proceeded in their governments, not only with an absolute power, but also with good order and iustice, suffering no man to be oppressed. The Ynca placed governours in divers Provinces, amongst the which some were superiors, and did acknowledge none but himselfe, others were of lesse commaund, and others more particular, with so goodly an order, and such gravitie, as no man durst bee drunke nor take an eare of Mays from his neighbour. These Yncas held it for a maxime, that it was necessary to keepe the Indians alwaies in action: and therefore we see it to this day, long cawseis and workes of great labour, the which they say were made to exercise the Indians, lest they should remaine idle. When he conquered any new Province, he was accustomed presently to send the greatest part, and the chiefe of that country into other Provinces, or else to his Court, and they call them at this day in Peru *Mitimas*, and in their places hee sent others of the Nation of Cusco, especially the Orejones, which were as Knights of an ancient house. They punished faultes rigorously. And therefore such as have any vnderstanding heereof hold opinion that there can be no better government for the Indians, nor more assured then that of the Yncas.

Chap. xiii.—*Of the distribution the Yncas made of their Vassals.*

To relate more particularly what I have spoken before, you must vnderstand that the distribution which the Yncas made of their vassals was so exact and distinct, as he might governe them all with great facilitie, although his realme were a thousand leagues long: for having conquered a Province, he presently reduced the Indians into Towns and Comminalties, the which he divided into bandes, hee appointed one to have the charge over every ten Indians, over every hundred another, over every thousand another, and over ten thousand another, whom they called Hunu, the which was one of the greatest charges. Yet above all in every Province, there was a Governour of the house of the Yncas, whom all the rest obeyed, giving vnto him every yeare in particular account of what had passed, that is, of such as were borne, of those that were dead, and of their troups and graine. The Governors went every yeare out of Cusco, where they remained, and returned to the great feast of Raymi, at the which they brought the tribute of the whole Realme to the Court; neither might they enter but with this condition. All the Kingdome was divided into foure partes, which they called Tahuantinsuyu, that is, Chinchasuyu, Collasuyu, Antisuyu, and Cuntisuyu, according to the foure waies which went from Cusco, where the Court was resident, and where the generall assemblies of the realme were made. These waies and Provinces being answerable vnto them, were towards the foure quarters of the world, Collasuyu to the South, Chinchasuyu to the North, Cuntisuyu to the West, and Antisuyu to the East. In every towne and village there were two sortes of people, which were of Hanansuyu and Urinsuyu, which is as much to say, as those above, and those below. When they commanded any worke to be done, or to furnish any thing to the Ynca, the officer knew presently how much every Province, Towne, and Family, ought to furnish, so as the division was not made by equall portions, but by cottization,[1] according to the qualities and wealth of the Countrie. So as for example, if they were to gather a hundred thousand Fanegas of Mays, they knew presently how much every Province was to contribute, were it a tenth, a seventh, or a fift part. The like was of Townes and Villages and Ayllus or Linages. The Quipocamayos, which were the officers and intendants, kept the account of all with their strings and knottes, without failing, setting downe what every one had paied, even to a hen, or a burthen of wood, and in a moment they did see by divers registers what every one ought to pay.

Chap. xiv.—*Of the Edifices and maner of building of the Yncas.*

The Edifices and Buildings which the Yncas made in temples, fortresses, waies, countrie houses, and such like, were many in number, and of an excessive labour, as doth appear at this day by their ruines and fragments that have remained, both in Cusco, Tiahuanaco,[2] Tambo,[3] and other places, where there are stones of an vnmeasurable greatnes, so as men cannot conceive how they were cut, brought, and set in their places. There came great numbers of people from all Provinces to worke in these buildings and fortresses, which the Ynca caused to be made in Cusco, or other partes of the Realme. As these workes were strange, and to amaze the beholders, wherein they vsed no mortar nor ciment, neither any yron, or steele, to cut, and set the stones in place. They had no engines or other instruments to carie them, and yet

[1] "Por quotas." [2] "Tiahuanaco." [3] "Ollantay-tampu."

were they so artificially wrought, that in many places they could not see the ioyntes, and many of these stones are so big, that it were an incredible thing if one should not see them. At Tiahuanaco I did measure a stone of thirty eight foote long, of eighteene broade, and six thicke. And in the wall of the fortresse of Cusco, which is of masonry,[1] there are stones of a greater bignes. And that which is most strange, these stones being not cut nor squared to ioyne, but contrariwise, very vnequall one with another in forme and greatnes, yet did they ioyne them together without ciment after an incredible maner. All this was done by the force of men who endured their labour with an invincible patience. For to ioyne one stone with an other, they were forced to handle and trie many of them often, being vneven. The Ynca appoynted every yeare what numbers of people should labour in these stones and buildings, and the Indians made a division amongest them, as of other things, so as no man was oppressed. Although these buildings were great, yet were they commonly ill appoynted and vnfit, almost like to the mosques or buildings of the Barbarians.

They could make no arches in their edifices, nor mortar or cyment to builde them withall. When they saw arches of wood built vpon the river of Xauxa, the bridge being finished, and the wood broken downe, they all beganne to runne away, supposing that the bridge, which was of stone, should presently fall; but when they found it to stand firme, and that the Spaniards went on it, the Cacique saide to his companions, "It is reason we should serve these men, who in trueth seeme to be the children of the Sunne". The bridges they made were of reedes plaited, which they tied to the bankes with great stakes, for that they could not make any bridges of stone or wood. The bridge which is at this day vpon the Desaguadero or river draining the great lake Chucuito[2] in Collao is admirable, for the course of that water

---

[1] "Mamposteria."   [2] Or Titicaca.

is so deep as they can not settle any foundation, and so broade
that it is impossible to make an arch to passe it, so as it
was altogether impossible to make a bridge eyther of wood
or stone. But the wit and industry of the Indians invented
a meanes to make a firme and assured bridge, being only
of strawe, which seemeth fabulous, yet is it very true. For
as we have said before, they did binde together certaine
bundles of reedes, and weedes, which do grow in the lake
that they call Totora, and being a light matter that sinkes
not in the water, they cast it vppon a great quantity of
reedes; then, having tied those bundles of weedes to either
side of the river, both men and beasts goe over it with
ease. Passing over this bridge I wondered, that of so
common and easie a thing, they had made a bridge, better,
and more assured than the bridge of boates from Seville
to Triana. I have measured the length of this bridge, and,
as I remember, it was above three hundred foote, and they
say that the depth of this current is very great; and it
seemes above, that the water hath no motion, yet they say,
that at the bottome it hath a violent and very furious
course. And this shall suffice for buildings.

CHAP. XV.—*Of the Yncas revenues, and the order of Tributes they imposed vpon the Indians.*

The Yncas riches was incomparable, for although no king
did inherite the riches and treasure of his predecessor,
yet had he at commaund all the riches of his realmes,
as well silver and gold, as the stuffe of Cumbi,[1] and cattell
wherein they abounded, and their greatest riches of all, was
their innumerable number of vassals, which were all
imployed as it pleased the King. They brought out of every
province what he had chosen for tribute. The Chichas sent
him sweete and rich woods; the Lucanas sent bearers to

[1] Fine cloth.

carry his Litter; the Chumbivilcas, dauncers; and so the other provinces sent him what they had of aboundaunce, besides their generall tribute, wherevnto every one contributed. The Indians that were appointed to that end, labored in the mines of golde and silver, which did abound in Peru, whom the Ynca intertained with all they needed for their expences; and whatsoever they drew of gold and silver, was for him. By this meanes there were so great treasures in this kingdome, as it is the opinion of many, that what fell in the handes of the Spaniardes, although it were very much, as wee know, was it not the tenth part of that which they hid and buried in the ground, the which they could never discover, notwithstanding all the search covetousnesse had taught them. But the greatest wealth of these barbarous people, was, that their vassalles were all slaves, whose labour they vsed at their pleasure; and that which is admirable, they imployed them in such sorte, as it was no servitude vnto them, but rather a pleasing life. But to vnderstand the order of tributes which the Indians payed vnto their Lordes, you must knowe, that when the Ynca conquered any citties, he divided all the land into three partes; the first was for religion and ceremonies, so as the Pachayachachi,[1] which is the Creator, and the Sunne, the Chuquilla, which is the Thunder, the Pachamama, and the dead, and other Guacas and sanctuaries, had every one their proper lands, the fruits whereof were spoyled and consumed in sacrifices, and in the nourishing of ministers and priests; for there were Indians appoynted for every Guaca, and sanctuary, and the greatest parte of this revenue was spent in Cusco, where was the vniversall and generall sanctuarie, and the rest in that cittie where it was gathered; for that after the imitation of Cusco, there were in every Citie, Guacas, and Oratories of the same order, and with the same functions, which were served after the same

[1] Teacher of the World.

manner and ceremonies to that of Cusco, which is an admirable thing, and they have found it by proofe in above a hundred townes, some of them distant above two hundred leagues from Cusco. That which they sowed or reapt vpon their land, was put into houses, as granaries, or store-houses, built for that effect, and this was a great parte of the Tribute which the Indians payed. I can not say how much this parte amounted vnto, for that it was greater in some partes than in other, and in some places it was in a manner all; and this parte was the first they put to profite. The second parte of these lands and inheritances was for the Ynca, wherewith he and his householde were entertained, with his kinsfolks, noblemen, garrisons and souldiers. And therefore it was the greatest portion of these tributes, as it appeareth by the quantity of golde, silver, and other tributes, which were in houses appoynted for that purpose, being longer and larger than those where they keepe the revenues of the Guacas. They brought this tribute very carefully to Cusco, or vnto such places where it was needefull for the souldiers, and when there was store, they kept it tenne or twelve yeares, vntill a time of necessitie. The Indians tilled and put to profite the Yncas lands, next to those of the Guacas; during which time they lived and were nourished at the charges of the Ynca, of the Sunne, or of the Guacas, according to the land they laboured. And the olde men, women, and sicke folkes were reserved and exempt from this tribute, and although whatsoever they gathered vpon those lands were for the Ynca, the Sunne, or the Guacas, yet the property appertayned vnto the Indians and their successors. The third parte of these landes were given by the Ynca for the comminaltie, and they have not yet discovered whether this portion were greater or lesse than that of the Ynca or Guacas. It is most certaine they had a care and regarde that it should be sufficient for the nourishment of the people. No particular man possessed any thing proper to himself of this third portion, neither did the Indians ever possesse any, if it were not by speciall grace from the Ynca; and yet might it not be engaged nor divided amongest his heires. They every yeare divided these landes of the comminaltie, in giving to every one that which was needful for the nourishment of their persons and families. And as the familie increased or diminished, so did they encrease or decrease his portion, for there were measures appoynted for every person. The Indians payed no tribute of that which was apportioned vnto them; for all their tribute was to till and keepe in good order the landes of the Ynca, and the Guacas, and to lay the fruits thereof in their store-houses. When the yeare was barren, they gave of these fruits thus reserved to the needy, for that there is alwayes superaboundance. The Ynca did likewise make distribution of the cattell as of the landes, which was to number and divide them; then to appoynt the pastures and limites, for the cattell belonging to the Guacas, and to the Ynca, and to everie Towne; and therefore one portion of their revenues was for religion, another for the Ynca, and the third for the Indians themselves. The like order was observed among the hunters, being forbidden to take or kill any females. The flocks of the Yncas and Guacas were in great numbers and very fruitfull; for this cause they called them Capacllama; but those of the common and publike, were few in number and of small valew, and therefore they called them Huacchallama.[1] The Ynca took great care for the preservation of cattell, for that it hath beene, and is yet, all the wealth of the Countrey, and as it is sayd, they did neither sacrifice any females, nor kill them, neither did they take them when they hunted. If the mange or the scurvie, which they call Carachi, take any beast, they were presently commaunded to bury it quicke, lest it should infect others. They did sheare their cattell in their season, and distributed to every

[9] *Ccapac*, rich; *Huaccha*, poor.

one to spinne and weave stuffes for the service of his familie. They had searchers to examine if they did employ themselves in these workes, and to punish the negligent. They made stuffes of the wooll of the Yncas cattell, for him and for his family, one sorte very fine, which they called Cumbi, and another grosser, which they likewise called Abasca.[1] There was no certaine number of these stuffes and garments appointed, but what was delivered to every one. The wooll that remayned was put into the storehouses, whereof the Spaniards found them ful, and with all other things necessary for the life of man. There are few men of iudgement but doe admire at so excellent and well settled a governement, seeing the Indians (being neyther religious, nor christians) maintained after their manner, this perfection, nor to holde any private property, and to provide for all necessities, also maintaining with such aboundance matters of religion, and that which concerned their King and Lord.

CHAP. XVI.—*Of arts and offices which the Indians did exercise.*

The Indians of Peru had one perfection, which was to teach their young children all artes and occupations necessary for the life of man; for that there were no particular trades-men, as amongst vs, taylors, shoemakers, weavers, and the rest, but everyone learned what was needefull for their persons and houses, and provided for themselves. All coulde weave and make their garments, and therefore the Ynca by furnishing them with wooll, gave them clothes. Every man could till the ground, and put it to profite, without hyring of any labourers. All built their owne houses, and the women vnderstoode most, they were not bred vppe in delights, but served their husbands carefully. Other arts and trades which were not ordinary and common for the life

[1] *Auasca*, coarse cloth.

of man, had their proper companies and workmen, as goldsmiths, painters, potters, watermen, and players of instruments. There were also weavers and workemen for exquisite workes, which the noblemen vsed: but the common people, as hath beene said, had in their houses all things necessary, having no need to buy. This continues to this day, so as they have no need one of another for things necessary: touching his person and family, as shoes and garments, and for their house, to sowe and reape, and to make yron woorkes, and necessary instruments. The Indians heerein doe imitate the institutions of the ancient monks, whereof is intreated in the lives of the Fathers. In trueth it is a people not greatly covetous, nor curious, so as they are contented to passe their time quietly, and without doubt, if they made choise of this manner of life, by election, and not by custome or nature, we may say that it was a life of great perfection, being apt to receive the doctrine of the holy Gospel, so contrary an enimy to pride, covetousness, and delights. But the preachers give not alwayes good example, according to the doctrine they preach to the Indians. It is woorthy observation, although the Indians be simple in their manner and habites, yet do wee see great diversitie amongest the provinces, especially in the attire of their head, for in some places they carried a long piece of cloth which went often about, in some places a large piece of cloth, which went but once about, in some parts as it were little morters or hattes, in some others as it were high and round bonets, and some like the bottome of sacks, with a thousand other differences. They had a straight and inviolable lawe, that no man might change the fashion of the garments of his province, although hee went to live in another. This the Ynca held to be of great importance for the order and good governement of his realme, and they doe observe it to this day, though not with so great a care as they were accustomed.

CHAP. XVII.—*Of the Posts and Chasquis the Indians did vse.*

There were many Posts and couriers which the Ynca maintained throughout his realme, whom they called Chasquis, and they carried commaundements to the Governours, and returned their advises and advertisements to the Court. These Chasquis were placed at every *topu*, which was a league and a halfe one from an other in two small houses, where were foure Indians. These were furnished by different districts, and changed monthly. Having received the packet or message, they ranne with all their force vntill they had delivered it to the other Chasquis, such as were to runne being ready and watchfull. They ran fifty leagues in a day and night, although the greatest parte of that countrey be very rough. They served also to carry such things as the Ynca desired to have with speede. Therefore they had always sea-fish in Cusco, of two dayes old or little more, although it were above a hundred leagues off. Since the Spaniardes entred, they have vsed of these Chasquis in time of seditions, whereof there was great need. Don Martin,[1] the Viceroy, appoynted ordinary posts at every foure leagues, to carry and recarry despatches, which were very necessary in this realme, though they run not so swiftly as the auntients did, neither are there so many, yet they are well payed, and serve as the ordinaries of Spaine, delivering letters, which they each carry foure or five leagues.

CHAP. XVIII.—*Of the iustice, lawes, and punishments which the Yncas have established, and of their marriages.*

Even as such as had done any good service in warre, or in the governement of the common-weale, were honoured

[1] Don Martin Henriquez.

and recompensed with publike charges, with lands given them in proper, with armes and titles of honour, and in marrying wives of the Yncas linage, even so they gave severe punishments to such as were disobedient and offenders. They punished murther, theft, and adultery, with death, and such as committed incest with ascendants or descendants in direct line, were likewise punished with death. But they held it no adultery to have many wives or concubines, neyther were the women subject to the punishment of death, being found with any other, but onely she that was the true and lawfull wife, with whome they contracted marriage; for they had but one whome they did wed and receive with a particular solempnitie and ceremony, which was in this maner: the bridegroome went to the bride's house, and led her from thence with him, having first put an *otoja*[1] vppon her foote. They call the shooe which they vse in those partes, *otoja*, being open like to the Franciscan Friars. If the bride were a mayde, her *otoja* was of wooll, but if she were not, it was of reedes. All his other wives and concubines did honour and serve this as the lawful wife, who alone, after the decease of her husband, caried a mourning weed of blacke, for the space of a yeare; neither did she marry vntil that time were past; and commonly she was yonger than her husband. The Ynca himselfe, with his own hand, gave this woman to his Governors and Captains; and the Governors or Caciques assembled all the young men and maydes, in one place of the City, where they gave to everyone his wife with the aforesaid ceremony, in putting on the *otoja*, and in this manner they contracted their marriages. If this woman were found with any other man than her husband, shee was punished with death, and the adulterer likewise: and although the husband pardoned them, yet were they punished, although dispensed withall from death. They

[1] *Usuta*, a sandal.

inflicted the like punishment on him that did commit incest with his mother, grandmother, daughter, or grandchilde : for it was not prohibited for them to marry together, or to have of their other kinsfolkes for concubines ; onely the first degree was forbidden. Neither did they allow the brother to have the company of his sister, wherein they of Peru were very much deceived, beleeving that the Yncas and noble men might lawfully contract marriage with their sisters, yea, by father and mother : for in trueth it hath beene alwayes helde vnlawfull among the Indians, and forbidden to contract in the first degree; which continued vntill the time of Tupac Ynca Yupanqui, father to Guaynacapa, and grandfather to Atahualpa, at such time as the Spaniards entered Peru, for that Tupac Ynca Yupanqui, was the first that brake this custome, marrying with Mamaocllo, his sister by the father's side, decreeing that the Yncas might marry with their sisters by the father's side, and no other.

This he did, and by that marriage he had Guaynacapa,[1] and a daughter called Coya Cusilimay. Finding himselfe at the poynt of death, he commaunded his children, by father and mother, to marry together, and gave permission to the noble men of his country, to marrie with their sisters by the fathers side. And for that this marriage was vnlawful, and against the lawe of nature, God would bring to an end this kingdome of the Ynca, during the raigne of Huascar Ynca, and Atahualpa Ynca, which was the fruite that sprang from this marriage. Whoso will more exactly vnderstand the manner of marriages among the Indians of Peru, lette him reade the treatise Polo hath written, at the request of Don Ieronimo Loaifa, Archbishop of the city of the Kings : which Polo made a very curious search, as he hath doone of divers other things at the Indies. The which importes much to be knowne to avoyde the errour and inconveniences where into

[1] Huayna Ccapac.

many fall (which know not which is the lawfull wife or the concubine among the Indians) causing the Indian that is baptized to marry with his concubine, leaving the lawfull wife: thereby also wee may see the small reason some have had, that pretended to say, that wee ought to ratifie the marriage of those that were baptized, although they were brother and sister. The contrary hath beene determined by the provinciall Synode of Lyma, with much reason, seeing among the Indians themselves this kind of marriage is vnlawful.

CHAP. XIX.—*Of the Originall of the Yncas, Lords of Peru, with their Conquests and Victories.*

By the commandement of Don Philip the Catholike King, they have made the most diligent and exact search that could be, of the beginning, customes, and priviledges of the Yncas, the which was not so perfectly done as was desired, for that the Indians had no written recordes; yet they have recovered that which I shall write by meanes of their Quippos and registers. First, there was not in Peru in olde time, any King or Lord to whome all obeyed, but they were comminalties, as at this day there be in the realme of Chile, and in a maner, in all the Provinces which the Spaniards have conquered in those westerne Indies, except the realme of Mexico. You must therefore understand that they have found three maner of governments at the Indies. The first and best was a Monarchie, as that of the Yncas, and of Moteçuma, although for the most part they were tyrannous. The second was of Comminalties, where they were governed by the advice and authoritie of many, which are as it were Counsellors. These in time of warre made choice of a Captaine, to whome a whole Nation or Province did obey; and in time of peace every Towne or Comminaltie did rule and governe themselves, having some chiefe men whom the vulgar did respect, and sometimes, though not often, some of them assemble together about matters of importance to consult what they should thinke necessary. The thirde kinde of government is altogether barbarous, composed of Indians without law, without King, and without any certaine place of abode, but go in troupes like savage beasts. As farre as I can conceive, the first inhabitants of the Indies were of this kinde, as at this day a great part of the Bresillians, Chiriguanas, Chunchos, Yscaycingas, Pilcoçones, and the greatest part of the Floridians, and all the Chichimecos in New Spaine. Of this kind the other sort of government by Comminalties was framed by the industrie and wisedome of some amongst them, in which there is some more order, holding a more staied place, as at this day those of Araucano, and of Tucapel in Chile, and in the new kingdome of Granada, the Moscas, and the Otomites in New Spaine; and in all these there is lesse fiercenes and incivilitie, and much more quiet then in the rest. Of this kinde, by the valure and knowledge of some excellent men, grew the other government more mightie and potent, which did institute a Kingdome and Monarchie. It appeares by their registers, that their government hath continued above three hundred yeares, but not fully foure, although their Seigniorie for a long time was not above five or six leagues compasse about the Citty of Cusco. Their originall and beginning was in the valley of Cusco, where by little and little they conquered the lands which we called Peru, passing beyond Quito, vnto the river of Pasto towardes the North, stretching even vnto Chile towardes the South, which is almost a thousand leagues in length. It extended in breadth vnto the South Sea towardes the west, and vnto the great champains which are on the other side of the Andes, where at this day is to be seene the castell which is called the Pucara of the Ynca, the which is a fortresse built for the defence of the frontire towards the East. The Yncas advanced no farther on that side, for the aboundance of water, marshes, lakes, and rivers, which runne in those partes. These Yncas passed all the other Nations of America in policy and government, and much more in valour and armes, although the Canaris which were their mortall enemies, and favoured the Spaniardes, would never confesse it, nor yeelde them this advantage; so as even at this day, if they fall into any discourse or comparisons, and that they be a little chafed and incensed, they kill one another by thousands vpon this quarrel, which are the most valiant, as it hath happened in Cusco. The practice and meanes which the Yncas had to make themselves Lords of all this Countrie, was in faining that since the generall deluge, whereof all the Indians have knowledge, the world had beene preserved, restored, and peopled by these Yncas, and that seven of them came foorth of the cave of Pacaritambo, by reason whereof, all other men owed them tribute and vassalage, as their progenitors. Besides, they said and affirmed, that they alone held the true religion, and knew how God should be served and honoured; and for this cause they should instruct all men. It is a strange thing the ground they give to their customes and ceremonies. There were in Cusco above foure hundred Oratories, as in a holy land, and all places were filled with their mysteries. As they continued in the conquests of Provinces, so they brought in the like ceremonies and customes. In all this realme the chiefe idol they did worship was Viracocha Pachayachachic,[1] which signifies the Creator of the world, and after him the Sunne. And therefore they said, that the Sunne received his vertue and being from the Creator, as the other idolls do, and that they were intercessors to him.

---

[1] Teacher of the world; from *Yachani*, I teach.

CHAP. XX.—*Of the first Ynca, and his Successors.*

The first man which the Indians report to be the beginning and first of the Yncas was Mangocapa,[1] whom they imagine, after the deluge, to have issued forth of the cave of Tambo, which is from Cusco about five or six leagues. They say that he gave beginning to two principall races or families of the Yncas, the one was called Hanancusco, and the other Vrincusco : of the first came the Lords which subdued and governed this Province, and the first whom they make the head and stem of this family was called Ingaroca,[2] who founded a family or Ayllu, as they call them, named Vicaquirao.[3] This, although he were no great Lord, was served notwithstanding in vessell of gold and silver. And dying, he appointed that all his treasure should be imployed for the service of his body, and for the feeding of his family. His successor did the like : and this grew to a generall custome, as I have said, that no Ynca might inherite the goods and house of his predecessor, but did build a new pallace. In the time of this Ingaroca the Indians had images of gold ; and to him succeeded Yaguarguaque,[4] a very old man : they say he was called by this name, which signifies teares of blood, for that being once vanquished and taken by his enemies, for griefe and sorrow he wept blood. He was buried in a village called Paulo, which is vpon the way to Omasuyo: he founded a family called Ayllu-panaca.[5] To him succeeded his sonne Viracocha Ynca, who was very rich and made much vessell of gold and silver: hee founded the linage or family of Cocopanaca. Gonzalo Pizarre sought out his body, for the report of the great treasure was buried with him, who, after he had cruelly

[1] Manco Ccapac.
[2] Ynca Rocca.
[3] Vicaquirau ; from *quirau*, a cradle.—See *G. de la Vega*, ii, p. 531.
[4] *Yahuar-huaccac*, literally, "Weeping blood".
[5] See *G. de la Vega*, ii, p. 531.

Lib. VI. tormented many Indians, in the end he found it in Xaquixaguana, whereas they said Pizarro was afterwards vanquished, taken, and executed by the President Gasca. Gonzalo Pizarro caused the body of Viracocha Ynca to be burnt; the Indians did afterwardes take the ashes, the which they preserved in a small vessell, making great sacrifices therevnto, vntill Polo did reforme it, and other idolatries which they committed vpon the bodies of their other Yncas, the which hee suppressed with an admirable diligence and dexterity, drawing these bodies out of their hands, being whole, and much imbalmed, whereby he extinguished a great number of idolatries which they committed. The Indians tooke it ill that the Ynca did intitle himselfe Viracocha, which is the name of their God: and he to excuse himselfe, gave them to vnderstand that the same Viracocha appeared to him in his dreame, commanding him to take this name. To him succeeded Pachacuti Ynca Yupanqui, who was a very valiant conquerour, a great politician, and an inventor of a great part of the traditions and superstitions of their idolatrie, as I will presently shew.

Chap. xxi.—*Of Pachacuti Ynca Yupanqui, and what happened in his time vnto Guaynacapa.*

Pachacuti Ynca Yupanqui reigned seventy yeares, and conquered many Countries. The beginning of his conquests was by meanes of his eldest brother, who, having held the government in his fathers time, and made warre by his consent, was over-throwne in a battle against the Chancas, a Nation which inhabites the valley of Andahuaylas thirty or forty leagues from Cusco, vpon the way to Lima. This elder brother thus defeated, retyred himselfe with few men. The which Ynca Yupanqui, his younger brother seeing, devised and gave forth that, being one day alone and melancholie, Viracocha, the Creator, spake to him, complaining that though he were vniversall Lord and Creator of all things, and that hee had made the heaven, the Sunne, the world, and men, and that all was vnder his command, yet did they not yeelde him the obedience they ought, but contrariwise did equally honour and worship the Sunne, Thunder, Earth, and other things, which had no virtue but what he imparted vnto them: giving him to vnderstand, that in heaven where hee was, they called him Viracocha Pachayachachic, which signifieth vniversall Creator; and to the end the Indians might beleeve it to be true, he doubted not although he were alone, to raise men vnder this title, which should give him victory against the Chancas, although they were then victorious, and in great numbers; and make himselfe Lord of those realmes, for that he would send him men to his aide invisibly, whereby he prevailed in such sort, that vnder this colour and conceit, hee beganne to assemble a great number of people, whereof he made a mighty armie, with the which he obtayned the victorie, making himselfe Lord of the whole Realme, taking the government from his father and brother. Then afterwardes he conquered and overthrew the Chancas, and from that time commanded that Viracocha should be held for vniversall Lord, and that the images of the Sunne and Thunder should do him reverence and honour. And from that time they beganne to set the image of Viracocha above that of the Sunne and Thunder, and the rest of the Guacas. And although this Ynca Yupanqui had given farmes, landes, and cattell to the Sunne, Thunder, and other Guacas, yet did he not dedicate any thing to Viracocha, saying that he had no neede, being vniversall Lord and Creator of all things. He informed his souldiers after this absolute victory over the Chancas, that it was not they alone that had conquered them, but certaine bearded men, whome Viracocha had sent him, and that no man might see them but himselfe, which were since converted into stones; it was therefore necessary to seeke them out whome he would know well. By this meanes hee gathered together a multitude of stones in the mountaines, whereof he made choice, placing them for Guacas, or Idolls, they worshipped and sacrificed vnto; they called them Pururaucas,[1] and carried them to the warre with great devotion, beleeving for certaine that they had gotten the victory by their help. The imagination and fiction of this Ynca was of such force, that by the means thereof hee obtained goodly victories. He founded the family called Ynacapanaca, and made a great image of golde, which hee called Ynti-yllapa, which hee placed in a brancard of golde, very rich, and of great price, of the which gold the Indians took great store to carry to Caxamarca for the libertie and ransome of Atahualpa, when the Marquis Francisco Pizarro held him prisoner. The Licentiate Polo found in his house in Cusco his servants and Mamaconas, which did service to his memorie, and found that the body had beene transported from Patallacta to Totocachi, where the Spaniards have since founded the parish of San Blas. This body was so whole and preserved with a certaine rosin, that it seemed alive; he had his eyes made of a fine cloth of golde, so artificially set, as they seemed very naturall eyes; he had a blowe with a stone on the head, which he had received in the warres; he was all grey and hairy, having lost no more haire than if hee had died but the same day, although it were seaventy and eight yeares since his decease. The foresaid Polo sent this body with some others of the Yncas to the cittie of Lima, by the viceroyes commaund, which was the Marquis of Cañete, and the which was very necessary to root out the idolatry of Cusco. Many Spaniards have seene this body with others in the hospital of San Andres, which the Marquis built, but they were much decayed. Don Felipe Caritopa, who was grand-child or

[1] See *G. de la Vega*, ii, p. 57.

great grand-childe to this Ynca, affirmed that the treasure he left to his family was great, which should be in the power of the Yanaconas, Amaru, Titu, and others. To this Ynca succeeded Tupac Ynca Yupanqui, to whom his son of the same name succeeded, who founded the family called Ccapac Sylla.[1]

Chap. xxii.—*Of the greatest and most famous Ynca called Guaynacapa.*

To this latter Ynca succeeded Guaynacapa, which is to say, a yoong man, rich and valiant,[2] and so was he in trueth more than any of his predecessors, or successors. Hee was very wise, planting good orders thorowout his whole realme, hee was a bold and resolute man, valiant, and very happy in warre. Hee therefore obtained great victories, and extended his dominions much farther then all his predecessors had done before him; he died in the realme of Quito, the which he had conquered, foure hundred leagues distant from his court. The Indians opened him after his decease, leauing his heart and entrailes in Quito; the body was carried to Cusco, the which was placed in the renowmed temple of the Sunne. We see yet to this day many cawseries, buildings, fortresses, and notable workes of this king: hee founded the familie of Tumi-bamba. This Guaynacapa was worshipped of his subjects for a god, being yet alive, as the olde men affirme, which was not doone to any of his predecessours. When he died, they slew a thousand persons of his householde, to serve him in the other life, all which died willingly for his service, insomuch that many of them offered themselves to death, besides such as were appoynted: his riches and treasure was admirable. And forasmuch as the

---

[1] See *G. de la Vega*, ii, p. 531.
[2] *Huayna*, young; *Ccapac*, rich.

Spaniards entred soone after his death, the Indians laboured much to conceale all, although a great parte thereof was carried to Caxamarca, for the ransome of Atahualpa, his sonne. Some woorthy of credite affirme that he hadde above three hundred sonnes and grand-children in Cusco. His mother, called Mamaocllo, was much esteemed amongst them. Polo sent her body, with that of Guaynacapa, very well imbalmed, to Lima, rooting out infinite idolatries. To Guaynacapa succeeded in Cusco, a sonne of his called Titu-cusi-hualpa, who since was called Huascar Ynca; his body was burned by the captaines of Atahualpa, who was likewise sonne to Guaynacapa, and rebelled in Quito against his brother, marching against him with a mighty armie. It happened that Quisquis and Chilicuchi, captains to Atahualpa, took Huascar Ynca in the cittie of Cusco, being received for Lord and king (for that hee was the lawfull successor) which caused great sorrowe throughout all his kingdome, especially in his Court. And as alwayes in their necessities they had recourse to sacrifices, finding themselves vnable to set their Lord at libertie, as well for the great power the captaines had that tooke him, as also, for the great army that came with Atahualpa, they resolved (some say by the commaundement of this Ynca) to make a great and solemne sacrifice to Viracocha Pachayachachic, which signifieth vniversall Creator, desiring him, that since they coulde not deliver their Lord, he would send men from heaven to deliver him from prison. And as they were in this great hope, vpon their sacrifice, news came to them, that a certaine people come by sea, was landed, and had taken Atahualpa prisoner. Heerevpon they called the Spaniards Viracochas, beleeving they were men sent from God, as well for the small number they were to take Atahualpa in Caxamarca, as also, for that it chaunced after their sacrifice done to Viracocha, and thereby they began to call the Spaniards Viracochas, as they doe at this day.

And in truth, if we had given them good example, and such as we ought, these Indians had well applied it, in saying they were men sent from God. It is a thing very well worthy of consideration, how the greatnesse and providence of God, disposed of the entry of our men at Peru, which had beene impossible, were not the dissention of the two brethren and their partisans, and the great opinion they hadde of christians, as of men sent from heaven, bound (by the taking of the Indians countrey) to labour to winne soules vnto Almightie God.

CHAP. XXIII.—*Of the last Successors Yncas.*

The rest of this subiect is handled at large by the Spanish Writers in the histories of the Indies, and for that it is not my purpose, I will speake only of the succession of the Yncas. Atahualpa being dead in Caxamarca, and Huascar in Cusco, and Francisco Pizarro with his people having seised on the realme, Mancocapa, sonne to Guaynacapa, besieged them in Cusco very straightly; but in the end he abandoned the whole countrey, and retired himselfe to Vilca-bamba, where he kept himselfe in the mountaines, by reason of the rough and difficult access, and there the successors Yncas remained, vntill Amaru, who was taken and executed in the market place of Cusco, to the Indians incredible griefe and sorrow, seeing iustice doone vpon him publiquely whome they helde for their Lorde.[1] After which time, they imprisoned others of the lineage of these Yncas. I have knowne Don Carlos, grand-childe to Guaynacapa, and son to Paullu, who was baptized, and alwayes favoured the Spaniards against Mancocapa his brother. When the Marquis of Cañete governed in this countrey, Sayri Tupac Ynca,

[1] Tupac Amaru, the last Ynca, was beheaded by order of the Viceroy Toledo in 1571.

went from Vilcabamba and came vpon assurance to the citty of Kings, where there was given to him the valley of Yucay, and other things, to whom succeeded a daughter of his. Beholde the succession which is knowne at this day of that great and rich familie of the Yncas, whose raigne continued above three hundred yeeres, wherein they reckon eleaven successors, vntill it was wholly extinguished. In the other linage of Vrincusco, which (as we have said before) had his beginning likewise from the first Mancocapa, they reckon eight successors in this sort. To Mancocapa succeeded Sinchi Rocca, to him Ccapac Yupanqui, to him Lloqui Yupanqui, to him Mayta Ccapac, to him Tarcoguaman, vnto whome succeeded his sonne, whome they name not, to this son succeeded Don Iuan Tambo, Maytapanaça. This sufficeth for the originall and succession of the Yncas, that governed the land of Peru, with that that I have spoken of their lawes, governement, and manner of life.

CHAP. XXIV.—*Of the manner of the Mexicaines common-weale.*

Although you may see by the historie which shall be written of the kingdome, succession, and beginning of the Mexicaines, their maner of commonweale and governement, yet will I speake briefly what I shall thinke fitte in generall to be most observed; whereof I will discourse more amply in the historie. The first point whereby we may iudge the Mexicaine governement to be very politike, is the order they had and kept inviolable in the election of their king; for since their first, called Acamapich, vnto their last, which was Monteçuma, the second of that name, there came none to the crowne by right of succession, but by a lawfull nomination and election. This election in the beginning was by

---

[1] This name is not in the lists of other authors.

the voyce of the commons, although the chiefe men managed it. Since in the time of Iscoatl the fourth king, by the advise and order of a wise and valiant man, called Tlacael, there were foure certayne Electours appoynted, which (with two lordes or kings subiect to the Mexicaine, the one of Tescuco and the other of Tacuba) had power to make this election. They did commonly choose yoong men for their kings, because they went alwayes to the warres, and this was in a manner the chiefe cause why they desired them so. They had a speciall regard that they shoulde be fit for the warres, and take delight and glory therein. After the election they made twoo kindes of feasts, the one in taking possession of the royall estate, for the which they went to the Temple, making great ceremonies and sacrifices vppon the harth, called Divine, where there was a continuall fire before the altare of the idoll, and after some Rhetoritians practised therein, made many orations and speeches. The other feast, and the most solemne, was at his coronation, for the which he must first overcome in battell, and bring a certaine number of captives, which they must sacrifice to their gods; he entred in triumph with great pompe, making him a solemne reception, as well they of the Temple, who went all in procession, sounding on sundry sortes of instruments, giving incense, and singing like secular men, as also the courtiers, who came forth with their devises to receive the victorious king. The Crowne or royall ensigne was before like to a Myter, and behinde it was cut, so as it was not round, for the fore parte was higher, and did rise like a poynt. The king of Tescuco had the privilege to crown the king of Mexico. The Mexicaines have beene very duetifull and loyall vnto their kings; and, it hath not beene knowne that they have practised any treason against them; onely their Histories report, that they sought to poison their king called Tiçocci, being a coward, and of small account; but it is not found that there hath beene any dissentions or partialities amongest them for ambition, thogh it be an ordinary thing in Comminalties; but contrariwise they reporte, as you shall see heereafter, that a man, the best of the Mexicaines, refused this realme, seeming vnto him to be very expedient for the Common-weale to have an other king. In the beginning, when the Mexicaines were but poore and weake, the kings were very moderate in their expenses and in their Court, but as they increased in power they increased likewise in pompe and state, vntill they came to the greatnesse of Monteçuma, who if hee had had no other thing but his house of beasts and birds, it had beene a prowde thing, the like whereof hath not beene seene; for there was in this house all sortes of fish, birds, and beasts, as in an other Noahs Arke, for sea fish there were pooles of salt-water, and for river fish lakes of freshwater, birds that do prey were fedde, and likewise wilde beasts in great aboundaunce; there were very many Indians imployed for the keeping of these beasts; and when he found an impossibilitie to nourish any sort of fish, fowle, or wilde beast, hee caused the image or likenesse to be made, richly cutte in pretious stones, silver, or golde, in marble, or in stone; and for all sortes of entertainements, hee had his severall houses and pallaces, some of pleasure, others of sorrowe and mourning, and others to treate of the affairs of the realme. There was in this pallace many chambers, according to the qualitie of noble men that served him, with a strange order and distinction.

CHAP. XXV.—*Of the titles and dignities the Indians vsed.*

The Mexicaines have beene very curious to divide the degrees and dignities amongst the Noble men and Lords, that they might distinguish them to whom they were to give the greatest honour. The dignity of these foure

Electors was the greatest, and most honourable next to the king, and they were chosen presently after the kings election. They were commonly brothers, or very neare kinsmen to the king, and were called Tlacohecalcatl, which signifies prince of darts, the which they cast, being a kind of armes they vse much. The next dignitie to this were those they doe call Tlacatecatl, which is to say circumcisers or cutters of men. The third dignitie were of those which they called Ezuahuacatl, which signifies a sheader of blood. Ali the which Titles and Dignities were exercised by men of warre. There was another, a fourth, intituled, Tlilancalqui, which is as much to say, as Lord of the blacke house, or of darkenesse, by reason of certaine incke wherewith the Priests annoynted themselves, and did serve in their idolatries. All these foure dignities were of the great Counsell, without whose advise the king might not doe anything of importance; and the king being dead they were to choose another in his place out of one of those foure dignities. Besides these, there were other Counsells and Audiences, and some say there were as many as in Spaine, and that there were divers seates and iurisdictions, with their Counsellers and Iudges of the Court, and others that were vnder them, as Corregidors, chiefe Iudges, captaines of Iustice, Lieutenants, and others, which were yet inferiour to these, with a very goodly order. All which depended on the foure first Princes that assisted the king. These foure onely had authority and power to condemne to death, and the rest sent them instructions of the sentences they had given. By meanes whereof they gave the king to vnderstand what had passed in his Realme.

There was a good order and settled policie for the revenues of the Crowne, for there were officers divided throughout all the provinces, as Receivers and Treasurers, which received the Tributes and royall revenews. And they carried the Tribute to the Court, at the least every moneth; which Tribute was of all things that doe growe or ingender on the land, or in the water, as well of iewells and apparrell, as of meat. They were very carefull for the well ordering of that which concerned their religion, superstition, and idolatries,: and for this occasion there were a great number of Ministers, to whom charge was given to teach the people the custome and ceremonies of their Lawe. Heerevppon one day a christian Priest made his complaint that the Indians were no good Christians, and did not profite in the lawe of God; an olde Indian answered him very well to the purpose in these terms: "Let the Priest, saide hee, imploy as much care and diligence to make the Indians christians, as the ministers of Idolles did to teach them their ceremonies; for with halfe that care they will make vs the best christians in the worlde, for that the lawe of Jesus Christ is much better; but the Indians learne it not, for want of men to instruct them." Wherein hee spake the very trueth, to our great shame and confusion.

CHAP. XXVI.—*How the Mexicaines made Warre, and of their Orders of Knighthood.*

The Mexicaines gave the first place of honour to the profession of armes, and therefore the Noblemen are their chiefe souldiers, and others that were not noble, by their valour and reputation gotten in warres, came to dignities and honours, so as they were held for noblemen. They gave goodly recompences to such as had done valiantly, who inioyed priviledges that none else might have, the which did much incourage them. Their armes were of rasors of sharpe cutting flints, which they set on either side of a staffe, which was so furious a weapon, as they affirmed that with one blow, they would cut off the necke of a horse. They had strange and heavy clubbes, lances fashioned like pikes, and other maner of dartes to cast, wherein they were very expert; but the greatest part of their combate was performed with stones. For defensive armes they had little rondaches or targets, and some kind of morions or head-pieces invironed with feathers. They were clad in the skinnes of tigres, lions, and other sauage beasts. They came presently to hands with the enemie, and were greatly practised to runne and wrestle, for their chief maner of combate, was not so much to kill, as to take captives, the which they vsed in their sacrifices, as hath beene said. Monteçuma set knighthood in his highest splendor, ordaining certaine militarie orders, as Commanders, with certaine markes and ensignes. The most honourable amongest the Knightes, were those that carried the crowne of their haire, tied with a little red ribband, having a rich plume of feathers, from the which, did hang branches of feathers vpon their shoulders, and roules of the same. They carried so many of these rowles, as they had done worthy deedes in warre. The King himselfe was of this order, as may be seene in Chapultepec, where Monteçuma and his sonnes were attyred with those kindes of feathers, cut in the rocke, the which is worthy the sight. There was another order of Knighthood, which they called the lions and the tigres, the which were commonly the most valiant and most noted in warre, they went alwaies with their markes and armories. There were other Knightes, as the grey Knightes, the which were not so much respected as the rest: they had their haire cut round about the eare. They went to the war with markes like to the other Knightes, yet they were not armed but to the girdle, and the most honourable were armed all over. All Knightes might carry golde and silver, and weare rich cotton, and use painted and gilt vessell, and carry shooes after their maner: but the common people might vse none but earthen vessell, neyther might they carry shooes, nor attyre themselves but in Nequen, the which is a grosse stuffe. Every order of these Knightes had his lodging in the pallace noted with their markes; the first was called the Princes lodging, the second of Eagles, the third of Lions and Tigres, and the fourth of the grey Knightes. The other common officers were lodged vnderneath in meaner lodgings: if any one lodged out of his place, he suffered death.

Chap. xxvii.—*Of the great order and diligence the Mexicaines vsed to instruct their youth.*

There is nothing that gives me more cause to admire, nor that I finde more worthy of commendations and memory, then the order and care the Mexicaines had to nourish their youth; for they knew well that all the good hope of a common weale consisted in the nurture and institution of youth, whereof Plato treates amply in his bookes *De Legibus;* and for this reason they laboured and tooke paines to sequester their children from delights and liberties, which are the two plagues of this age, imploying them in honest and profitable exercises. For this cause there was in their Temples a private house for childeren, as schooles, or colledges, which was seperate from that of the yong men and maides of the Temple, whereof we have discoursed at large. There were in these schooles a great number of children, whom their fathers did willingly bring thither, and which had teachers and masters to instruct them in all commendable exercises, to be of good behaviour, to respect their superiors, to serve and obey them, giving them to this end certain precepts and instructions. And to the end they might be pleasing to Noblemen, they taught them to sing and dance, and did practise them in the exercise of warre, some to shoote an arrow, to cast a dart or a staffe burnt at the end, and to handle well a target and a sword. They suffered them not to sleepe much, to the end they might accustome themselves to labour in their youth, and were not men given to delights. Besides the ordinary number of these children, there were in the same colledges other children of Lordes and Noblemen, the which were instructed more privately. They brought them their meate and ordinary from their houses, and were recommended to antients and old men to have care over them, who continually did advise them to be vertuous and to live chastely; to be sober in their diet, to fast, and to march gravely, and with measure. They were accustomed to exercise them to travell, and in laborious exercises; and when they see them instructed in all these things, they did carefully looke into their inclination, if they found any one addicted to the war, being of sufficient yeares, they sought all occasions to make triall of them, sending them to the warre, vnder colour to carry victualls and munition to the souldiers, to the end they might there see what passed, and the labour they suffered. And that they might abandon all feare, they were laden with heavy burthens, that shewing their courage therein, they might more easily be admitted into the company of souldiers. By this meanes it happened that many went laden to the Armie and returned Captaines with markes of honour. Some of them were so desirous to bee noted, as they were eyther taken or slaine; and they held it lesse honourable to remaine a prisoner; and, therefore, they sought rather to be cut in peeces then to fall captives into their enemies hands. See how Noblemens children that were inclined to the warres were imployed. The others that had their inclination to matters of the Temple; and to speake after our maner, to be Ecclesiastical men, having attained to sufficient yeares, they were drawne out of the colledge, and placed in the temple in the lodging appointed for religious men, and then they gave them the orders of Ecclesiasticall men. There had they prelates and masters to teach them that which concerned their profession, where they should remaine being destined therevnto. These Mexicaines tooke great care to bring vp their children: if at this day they would follow this order, in building of houses and colledges for the instruction of youth, without doubt Christianitie should florish much amongst the Indians. Some godly persons have begunne, and the King with his Counsell have favored it: but for that it is a matter of no profit, they advance little, and proceed coldly. God open our eyes, that we may see it to our shame, seeing that we Christians do not that which the children of darkenes did to their perdition, wherin we forget our duties.

CHAP. XXVIII.—*Of the Indians feasts and dances.*

Forasmuch as it is a thing which partly dependes of the good government of the Common-weale, to have some plaies and recreations when time serves; it shall not be from the purpose to relate what the Indians did heerein, especially the Mexicaines. We have not discovered any Nation at the Indies that live in commonalties, which have not their recreations in plaies, dances, and exercises of pleasure. At Peru I have seene plaies in maner of combats, where the men of both sides were sometimes so chafed that often their *Puclla* (which was the name of this exercise) fell out to be dangerous. I have also seene divers sortes of dances, wherein they did counterfait and represent certaine trades and offices, as sheepherds, laborers, fishers, and hunters, and commonly they made all those dances with a very grave sound and pase: there were other dances and maskes, which they called cuacones, whose actions were pure representations of the divell. There were also men that dance on the shoulders one of another, as they do in Portugall, the which they call *pelas*. The greatest part of these dances were superstitions and kindes of idolatries: for that they honoured their idolls and Guacas in that maner. For this reason the Prelates have laboured to take from them these dances all they could: but yet they suffer them, for that part of them are but sportes of recreation, for alwaies they dance after their maner. In these dances they vse sundry sortes of instruments, whereof some are like flutes or little lutes, others like drummes, and others like shells: but commonly they sing all with the voyce, and first one or two sing the song, then all the rest answer them. Some of these songs were very wittily composed, contayning histories, and others were full of superstitions, and some were meere follies. Our men that have conversed among them have laboured to reduce matters of our holy faith to their tunes, the which hath profited well: for that they imploy whole daies to rehearse and sing them, for the great pleasure and content they take in their tunes. They have likewise put our compositions of musicke into their language, as Octaves, Songs, and Rondells, the which they have very aptly turned, and in truth it is a goodly and very necessary meanes to instruct the people. In Peru they commonly called dances *Taqui*, in other Provinces *Areytos*, in Mexico *Mitotes*. There hath not beene in any other place any such curiositie of plaies and dances as in New Spaine, where at this day we see Indians so excellent dancers, as it is admirable. Some dance vpon a cord, some vpon a long and straight stake, in a thousand sundrie sortes, others with the soles of their feete and their hammes do handle, cast vp, and receive againe a very heavy blocke, which seems incredible but in seeing it. They do make many other shewes of their great agilitie in leaping, vaulting, and tumbling, sometimes bearing a great and heavie burthen, sometimes enduring blowes able to breake a barre of yron. But the most usuall exercise of recreation among the Mexicaines is the solemne *Mitote*, and that is a kinde of daunce they held so brave and so honorable, that the king himselfe daunced, but not ordinarily, as the king Don Pedro of Aragon with the Barber of Valencia. This daunce or *Mitote* was commonly made in the Courts of the Temple, and in those of the kings houses, which were more spatious. They did place in the midst of the Court two instruments, one like to a drumme, and the other like a barrell made of one peece, and hollow within, which they set vppon the forme of a man, a beast, or vpon a piller.

These two instruments were so well accorded together, that they made a good harmony : and with these instruments they made many kinds of aires and songs. They did all sing and dance to the sound and measure of these instruments, with so goodly an order and accord, both of their feete and voices, as it was a pleasant thing to beholde. In these daunces they made two circles or wheeles, the one was in the middest neere to the instruments, wherein the Auntients and Noblemen did sing and daunce with a softe and slowe motion ; and the other was of the rest of the people round about them, but a good distance from the first, wherein they daunced two and two more lightly, making diverse kindes of pases, with certaine leapes to the measure. All which together made a very great circle. They attired themselves for these dances with their most pretious apparell and iewelles, every one according to his abilitie, holding it for a very honorable thing : for this cause they learned these daunces from their infancie. And although the greatest parte of them were doone in honor of their Idolles, yet was it not so instituted, as hath bin said, but only as a recreation and pastime for the people. Therefore it is not convenient to take them quite from the Indians, but they must take good heed they mingle not their superstitions amongest them. I have seene this *Mitote*, in the court of the Church of Tepotzotlan, a village seven leagues from Mexico : and, in my opinion, it was a good thing to busie the Indians vpon festivall dayes, seeing they have neede of some recreation : and because it is publike, and without the prejudice of any other, there is lesse inconvenience than in others, which may be done privately by themselves, if they tooke away these. We must therefore conclude, following the counsel of pope Gregory, that it was very convenient to leave vnto the Indians that which they had usually of custom, so as they be not mingled nor corrupt with their antient errors, and that their feasts and pastimes may be to the honor of God and of the Saints, whose feasts they celebrate. This may suffice in generall of the maners and politike customes of the Mexicaines. And as for their beginning, increase, and Empire, for that it is an ample matter, and will be pleasant to vnderstand from the beginning, we will intreate thereof in the Booke following.

THE SEVENTH BOOKE

Of the Naturall and Morall Historie of the Indies.

CHAP. I.—*That it is profitable to vnderstand the actes of the Indians, especially of the Mexicaines.*

LIB. VII.
Eccles. i.

EVERY History, wel written, is profitable to the reader: For as the Wise man saith, "That which hath bin, is, and that which shall be, is that which hath beene." Humane things have much resemblance in themselves, and some growe wise by that which happeneth to others. There is no Nation, how barbarous so ever, that have not something in them good, and woorthy of commendation; nor Commonweale so well ordered, that hath not something blameworthy, and to be controlled. If, therefore, there were no other fruite in the Historie and Narration of the deedes of the Indians, but this common vtilitie, to be a Relation or Historie of things, the which in the effect of truth have happened, it deserveth to be received as a profitable thing, neither ought it to be reiected, for that it concernes the Indians. As we see that those Authors that treate of naturall things, write not onely of generous beasts, notable and rare plants, and of pretious stones, but also of wilde beasts, common hearbes, and base and vulgar stones, for that there is always in them some properties worthy observation. If, therefore, there were nothing else in this Discourse, but that it is a Historie, and no fables nor fictions, it were no vnwoorthy subject to be written or read.

There is yet an other more particular reason, which is, that wee ought heerin to esteeme that which is woorthy of memorie, both for that it is a Nation little esteemed, and also a subiect different from that of our Europe, as these Nations be, wherein wee should take most pleasure and content, to vnderstand the ground of their beginning, their maner of life, with their happy and vnhappy adventures. And this subiect is not onely pleasaut and agreeable, but also profitable, especially to such as have the charge to rule and governe them; for the knowledge of their acts invites vs to give credite, and dooth partely teach howe they ought to be intreated : yea, it takes away much of that common and foolish contempt wherein they of Europe holde them, supposing that those Nations have no feeling of reason. For in trueth wee can not cleere this errour better, than by the true report of the actes and deedes of this people. I will, therefore, as briefly as I can, intreate of the beginning, proceedings, and notable deedes of the Mexicaines, whereby wee may know the time and the disposition that the high God woulde choose, to send vnto these Nations the light of the Gospel of Iesus Christ his only sonne our Lord, whome I beseech to second our small labour, that it may be to the glory of his Divine greatnes, and some profite to these people, to whome hee hath imparted the lawe of his holy gospel.

CHAP. II.—*Of the ancient Inhabitants of New Spaine, and how the Navatlacas came thither.*

The antient and first Inhabitants of those provinces, which wee call New Spaine, were men very barbarous and savage, which lived onely by hunting, for this reason they were called Chichimecas. They did neither sowe nor till the ground, neither lived they together; for all their exercise was to hunt, wherein they were very expert. They lived in the roughest partes of the mountaines beastlike, without any pollicie, and they went all naked. They hunted wilde beasts, hares, connies, weezles, mowles, wilde cattes, and birdes, yea vncleane beasts, as snakes, lizards, locusts, and wormes, whereon they fed, with some hearbs and rootes. They slept in the mountaines, in caves and in bushes, and the wives likewise went a hunting with their husbandes, leaving their yoong children in a little panier of reeds, tied to the boughs of a tree, which desired not to suck vntill they were returned from hunting. They had no superiors, nor did acknowledge or worship any gods, neyther hadde any manner of ceremonies or religion.

There is yet to this day in New Spaine of this kinde of people, which live by their bowes and arrowes, the which are very hurtfull, for that they gather together in troupes to doe mischiefe, and to robbe: neither can the Spaniards by force or cunning reduce them to any pollicie or obedience: for having no towns nor places of residence, to fight with them, were properly to hunt after savage beasts, which scatter and hide themselves in the most rough and covered places of the mountaines. Such is their maner of living even to this day, in many Provinces of the Indies. In the Bookes *De procuranda Indorum salute*, they discourse chiefly of this sort of Indians, where it is saide that they are to be constrained and subiected by some honest force, and that it is necessary first to teach them that they are men, and then to be Christians. Some will say that those in New Spaine, which they call Otomies, were of this sort, being commonly poore Indians, inhabiting a rough and barren land, and yet they are in good numbers, and live together with some order, and such as do know them, find them no lesse apt and capable of matters of Christian religion, than others which are held to be more rich and better governed. Comming, therefore, to our subiect, the

Chichimecas and Otomies, which were the first inhabitants of New Spaine, for that they did neyther till nor sowe the land, they left the best and most fertile of the country vnpeopled, which Nations that came from farre did possess, whome they called Navatlacas, for that it was a more civill and pollitike Nation; this word signifies a people that speakes well, in respect of other barbarous nations without reason. These second peoplers, Navatlacas, came from other farre countries, which lie toward the north, where now they have discovered a kingdome they call New Mexico.

There are two provinces in this countrey, the one called Aztlan, which is to say a place of Herons: the other Tuculhuacan, which signifies a land of such, whose grandfathers were divine. The Inhabitants of these provinces have their houses, their lands tilled, gods, customes, and ceremonies, with like order and governement to the Navatlacas, and are divided into seven Tribes or Nations: and for that they have a custome in this province, that every one of these lineages hath his place and private territory. The Navatlacas paint their beginning and first territory in figure of a cave, and say that they came forth of seven caves to come and people the land of Mexico, whereof they make mention in their Historie, where they paint seven caves and men comming forth of them. By the computation of their bookes, it is above eight hundred yeeres since these Navatlacas came foorth of their country, reducing which to our accompt, was about the yeere of our Lord 720, when they left their country to come to Mexico, they stayed foure score years vpon the way; and the cause of this their long stay in their voyage, was, that their gods (which without doubt were divells, and spake visibly vnto them) had perswaded them to seeke new lands that had certaine signes. And therefore they came discovering the whole land, to search for these tokens which their Idolls had given them; and in places where they found any good dwellings, they peopled it, and laboured the land, and as they discovered better countries, they left those which they had first peopled, leaving still some, especially the aged, sick folkes, and the weary; yea, they did plant and build there, whereof we see the remainders at this day. In the way where they passed, they spent fourescore yeares in this manner of leasurely travell, the which they might have done in a moneth. By this meanes they entred the land of Mexico in the yeare nine hundred and two, after our computation.

Chap. III.—*How the six Lineages of Navatlacas peopled the land of Mexico.*

These seven Lineages I have spoken of, came not forth all together: the first were the Suchimilcos, which signifie a Nation of the seedes of flowers. Those peopled the bankes of the great lake of Mexico towards the South, and did build a cittie of their name, and many villages. Long time after came they of the second lineage called Chalcas, which signifies people of mouthes, who also built a cittie of their name, dividing their limmits and territories with the Suchimilcos. The third were the Tepanecas, which signifies people of the bridge: they did inhabit vpon the banke of the lake towards the West, and they increased so, as they called the chiefe and Metropolitane of their Province, Azcapuzalco, which is to say, an Ants nest, and they continued long time mighty. After them came those that peopled Tezcuco, which be those of Culhua, which is to say, a crooked people: for that in their Countrey there was a mountaine much bending.[1] And in this sort this lake was invironed with these foure Nations, these inhabiting on the East, and the Tepanecas on the North. These of Tezcuco, were held for great Courtiers, for their tongue and pronuntiation is very sweete and pleasant. Then arrived the

[1] "Cerro muy encorvado."

Tlatluicas, which signifies men of the Sierra or mountaine. Those were the most rude and grosse of all the rest, who finding all the plaines about the lake possessed even vnto the Sierra, they passed to the other side of the mountaine, where they found a very fertile, spatious and warme countrey, where they built many great villages, calling the Metropolitane of their province, Quahunahuac, which is as much to say, as a place that sounds the voice of an Eagle, which our common people call by corruption, Quernavaca, and at this day they call this province the Marquisate. Those of the sixt generation, which are the Tlascaltecas, which is to say men of bread, passed the mountaine towards the east, crossing all the Sierra Nevada, where that famous Vulcan is betwixt Mexico and the Ciudad de los Angeles, where they did finde a good country, making many buildings. They built many townes and citties, whereof the Metropolitane was called by their name Tlascala. This is the nation which favoured the Spaniards at their entrie, by whose help they did winne this country, and therefore to this day they pay no tribute but enioy a generall exemption. When all these Nations peopled these countries, the Chichimecas being the antient inhabitants, made no resistance, but fledde, and as people amazed they hid themselves in the most obscure of the rockes. But those that inhabited on th' other side of the mountaine where the Tlascaltecas had planted themselves, did not suffer them in quiet, as the rest of the Chichimecas had done, but they put themselves in defence to preserve their country, and being giants, as the Histories report, they sought to expell the last comers, but they were vanquished by the policy of the Tlascaltecas, who counterfeiting a peace with them, they invited them to a great banquet, and when they were busiest in their drunkennes, there were some laide in ambush, who secretly stole away their weapons, which were great clubbes, targets, swords of wood, and other such armes. Then did they sodainely set vpon them, and the Chichimecas seeking to defend themselves, they did want their armes, so as they fled to the mountaines and forrests adioyning, where they pulled downe trees as if they had beene stalkes of lettices. But, in the end, the Tlascaltecas being armed, and marching in order, they defeated all the giants, not leaving one alive. We must not holde this of the giants to be strange or a fable; for, at this day, we finde dead mens bones of an incredible bignes.

When I was in Mexico, in the yeare of our Lorde one thousand five hundred eighty sixe, they found one of those giants buried in one of our farmes, which we call Iesus del Monte, of whom they brought a tooth to be seene, which (without augmenting) was as big as the fist of a man; and, according to this, all the rest was proportionable, which I saw and admired at his deformed greatnes. The Tlascaltecas, by this victory, remained peaceable, and so did the rest of the lineages. The six lineages did alwayes entertaine amitie together, marrying their children one with another, and dividing their limites quietly: then they studied with an emulation to encrease and beautifie their common-weale. The barbarous Chichimecas, seeing what passed, beganne to vse some government, and to apparrell themselves, being ashamed of what had passed: for till then they had no shame. And having abandoned feare by their communication with these other people, they beganne to learne many things of them, building small cottages, having some pollicie and government. They did also choose Lordes, whom they did acknowledge for their superiors, by meanes whereof they did in a manner quite abandon this brutish life, yet did they alwayes continue in the Mountaines divided from the rest.

Notwithstanding, I hold it for certaine that this feare hath growne from other Nations and Provinces of the Indies, who at the first were savage men, who living onely by hunting, piercing the rockie and rough countries, discovering a new world, the inhabitants whereof were almost like savage beasts, without coverings or houses, without tilled landes, without cattell, without King, Law, God, or Reason. Since others, seeking better and new lands, inhabited this fertile Countrey, planting pollitike order and a kinde of common-weale, although it were very barbarous. After the same men, or other Nations, that had more vnderstanding then the rest, laboured to subdue and oppresse the lesse mighty, establishing Realmes and great Empires. So it happened in Mexico, at Peru, and in some partes where they finde Citties and Common-weales planted among these Barbarians. That which confirmes me in my opinion (whereof I have amply discoursed in the first booke), that the first inhabitants of the West Indies came by land, and so by consequence that the first continent of the Indies ioynes with that of Asia, Europe, and Affrike, and the new world with the old, although they have not yet discovered any countrey that toucheth and ioynes with the other world; or if there be any sea betwixt the two, it is so narrow that wilde beasts may easily swim over, and men in small boates. But leaving this Philosophie, let vs returne to our history.

CHAP. IV.—*Of the Mexicaines departure, of their iourney and peopling the Province of Mechoacan.*

Three hundred and two yeares after, the former two lineages had left their Country to inhabite New Spaine, the Country being now well peopled and reduced to some forme of government. Those of the seventh cave or line arrived, which is the Mexicaine Nation, the which, like vnto the rest, left the Province of Aztlan and Teuculhuacan, a pollitike, courtlike, and warlike Nation. They did worship the Idoll Vitzilipuztli, whereof ample mention hath beene made, and the divell that was in this idoll spake, and governed this

Nation easily. This idoll commanded them to leave their Country, promising to make them Princes and Lords over all the Provinces which the other six Nations did possesse, that hee would give them a land abounding with gold, silver, pretious stones, feathers, and rich mantells: wherevpon they went forth, carrying their idoll with them in a coffer of reedes, supported by foure of their principall priests, with whome he did talke and reveale vnto them in secret, the successe of their way and voyage, advising them of what should happen. He likewise gave them lawes, and taught them the customes, ceremonies, and sacrifices they should observe. They did not advance nor moove without commandement from this idoll. He gave them notice when to march and when to stay in any place, wherein they wholy obeyed him. The first thing they did wheresoever they came was to build a house or tabernacle for their false god, which they set alwaies in the middest of their Campe, and there placed the Arke vppon an altare, in the same manner as they have vsed in the holy Christian Church. This done, they sowed their land for bread and pulses, which they vsed: and they were so addicted to the obedience of their god, that if he commanded them to gather, they gathered; but if he commanded them to raise their campe, all was left there for the nourishment of the aged, sicke, and wearie, which they left purposely from place to place, that they might people it, pretending by this meanes that all the land should remaine inhabited by their Nation. This going forth and peregrination of the Mexicaines will happily seeme like to that of Egypt, and to the way which the children of Israell made, seeing that they, as well as those, were warned to go forth and to seeke the land of promise, and both the one and the other carried their god for their guide, consulted with the arke and made him a tabernacle, and he advised them, giving them lawes and ceremonies, and both the one and the other spake many yeares in their voyage to their promised land, where we observe the resemblance of many other things, as the histories of the Mexicaines do report, and the holy scriptures testifie of the Israelites. And without doubt it is a true thing, that the Divell, the prince of pride, hath laboured by the superstitions of this Nation, to counterfaite and imitate that which the most high God did with this Nation: for, as is said before, Satan hath a strange desire to compare and make himselfe equal with God: so as this mortall enemy hath pretended falsely to vsurpe what communication and familiaritie he hath pleased with men. Was there ever divell found so familiarly conversant with men as this divell Vitzilipuztli. We may wel iudge what he was, for that there was never seene nor heard speake of customes more superstitious, nor sacrifices more cruel and inhumane, than those which he taught them. To conclude, they were invented by the enemy of mankinde. The chiefe and Captaine whome they followed was called Mexi, whence came the name of Mexico, and of the Mexicaine Nation. This people marching thus at leisure, as the other six Nations had done, peopling and tilling the land in divers partes, whereof there is yet some shewes and ruines: and after they had endured many travells and dangers, in the end they came to the Province of Mechoacan, which is as much to say, a land of fish, for there is great abundance in goodly great lakes, where, contenting themselves with the situation and temperature of the ground, they resolved to stay there. Yet, having consulted with their idoll vpon this point, and finding him vnwilling, they demanded license to leave some of their men to people so good a land, the which he granted, teaching them the meanes how to do it, which was, that when the men and women should be entred into a goodly lake called Pazcuaro to bathe themselves, those which remained on land should steale away all their clothes, and then secretly raise their campe and depart without any bruite, the which was effected, and the rest which dreamt not of this deceit (for the pleasure they tooke in bathing), comming forth and finding themselves spoiled of their garments, and thus mocked and left by their companions, they remained discontented and vexed therewith : so as, to make shew of the hatred they had conceived against them, they say that they changed their maner of life and their language. At the least, it is most certaine that the Mechoacans have been alwaies enemies to the Mexicaines, and therefore they came to congratulate the Marquis Del Valle,[1] after his victory obtained when he had conquered Mexico.

CHAP. V.—*Of that which happened in Malinalco, Tula, and in Chapultepec.*

From Mechoacan to Mexico are above fifty leagues, and vpon the way is Malinalco, where it happened that complaining to their idoll of a woman that was a notable witch, which came in their company carrying the name of a sister to their god, for that with her wicked artes she did them much harme, pretending by certaine meanes to be worshipped of them as their goddesse: the idoll spake in a dreame to one of those old men that carried the arke, commaunding him to comfort the people, making them new and great promises, and that they should leave this his sister with her family, being cruell and bad, raising their campe at mid-night in great silence, leaving no shew what way they passed. So they did, and the witch remaining alone with her family, in this sort peopled a towne which they call Malinalco, the inhabitants whereof are held for great sorcerers, being issued from such a mother. The Mexicaines, for that they were greatly diminished by these divisions, and by the number of sicke and wearied persons which they had left behind, meant to repaire themselves,

[1] Hernan Cortes.

and to stay in a place called Tula, which signifies a place of reedes. There their idoll commanded them to stoppe a great river, that it might cover a great plaine, and by the meanes he taught them they did inviron a little hill called Coatepec, making a great lake, the which they did plant round about with willows, elmes, sapines, and other trees. There beganne to breede much fish, and many birdes came thither: so as it became a very pleasant place. The situation of this place seeming pleasant vnto them, and being wearied with travell, many talked of peopling there, and to passe no farther: wherewith the divell was much displeased, threatning the priests with death, commanding them to returne the river to hir course, saying that he would that night chastise those which had beene disobedient as they had deserved. And as to do ill is proper to the Divell, and that the divine Iustice doth often suffer such to be delivered into the hands of such a tormentor, that choose him for their god; it chanced that about mid-night they heard a great noise in one part of the campe, and in the morning going thither they found those dead that had talked of staying there. The maner of their death was, that their stomackes were opened and their hearts pulled out. And by that meanes this good god taught these poore miserable creatures the kindes of sacrifices that pleased him, which was in opening the stomacke to pull out the heart, as they have since practised in their horrible sacrifices. Seeing this punishment, and that the plaine was dried, the lake being emptied, they asked counsell of their god what to doe, who commanded them to passe on, the which they did by little and little, vntill they came to Chapultepec, a league from Mexico, famous for the pleasantnes thereof. They did fortifie themselves in these mountaines, fearing the nations which inhabited that Country, the which were opposite vnto them, especially for that one named Copil, sonne to this sorceresse, left in Malinalco, had blamed and spoken ill of the Mexicaines: for this Copil, by the commandement of his mother, awhile after followed the Mexicaines course, labouring to incense the Tepanecas and other neighbours against them, even vnto the Chalcas: so as they came with a strong army to destroy the Mexicaines. Copil, in the meane space, stoode vpon a little hill in the middest of a lake called Acopilco, attending the destruction of his enemies, and they, by the advise of their idoll, went against him, tooke him suddenly, and slew him, carrying his heart to their god, who commanded them to cast it into the lake, faining that thereof did grow a plant called Tunal,[1] where since Mexico was built. They came to fight with the Chalcas and other Nations, having chosen for their Captaine a valiant man called Vitzilonitli, who, in an encounter, was taken and slaine by the enemies. But for all this, they were not discouraged, but fought valiantly; and in dispight of their enemies they brake the squadrons, and carrying their aged, their women, and yong children in the midst of their battaile, they passed on to Atlacuyavaya, a town of the Culhuas, whom they found solemnising of a feast, in which place they fortified. The Chalcas, nor the other Nations, did not follow them, but grieved to be defeated by so small a number of men; they being in so great multitudes retyred to their townes.

CHAP. VI.—*Of the Warres the Mexicaines had against them of Culhuacan.*

The Mexicaines, by the advice of their idoll, sent their messengers to the Lord of Culhuacan, to demand a place to dwell in, who after he had imparted it to his people, granted them the place of Tiçaapan, which signifies white waters, to the end they should all perish there, being full of vipers, snakes, and other venomous beasts which bred in a

[1] Prickly pear.

hill neere adioyning. But being perswaded and taught by their divell, they accepted willingly what was offered, and by their divelish art tamed these beastes, so as they did them no harme; yea, they vsed them as meat, eating them with delight and appetite. The which the Lord of Culhuacan seeing, and that they had tilled and sowed the land, he resolved to receive them into the Cittie, and to contract amity with them. But the god whom the Mexicaines did worship (as he is accustomed to do no good, but ill) said vnto his priests, that this was not the place where he would have them stay, and that they must go forth making warres. Therefore they must seeke forth a woman, and name her the goddesse of Discord. Wherevpon they resolved to send to the King of Culhuacan, to demand his daughter to be Queene of the Mexicaines, and mother to their god, who received this Ambassage willingly, sending his daughter presently gorgeously attyred and well accompanied. The same night she arrived, by order of the murtherer whome they worshipped, they killed her cruelly, and having flaed her artificially as they could do, they did clothe a yong man with her skinne, and therevpon her apparrell, placing him neere their idoll, dedicating him for a goddesse and the mother of their god, and ever after did worship it, making an idoll which they called Tocci, which is to say our grandmother. Not content with this crueltie, they did maliciously invite the King of Culhuacan, the father of the yong maid, to come and worshippe his daughter, who was now consecrated a goddesse, who comming with great presents, and well accompanied with his people, he was led into a very darke chappell where their idoll was, that he might offer sacrifice to his daughter that was in that place. But it chanced that the incense that was vpon the harth, according to their custome, kindled in such sort, as hee might discerne his daughter's haire, and having by this meanes discovered the cruelty and deceit, hee went forth crying alowde, and with all his men he fell vpon the Mexicaines, forcing them to retyre to the lake, so as they were almost drowned. The Mexicaines defended themselves, casting certaine little darts, which they vsed in the warres, wherewith they much galled their ennemies. But in the end they got land, and leaving that place, they coasted along the lake, very weary and wet; the women and little children crying and making great exclamations against them and their god that had brought them into this distresse. They were inforced to passe a river that could not be waded through, and therefore they advised to make small boates of their targets, and of reedes, wherein they passed. Then afterwardes, having left Culhuacan, they arrived at Iztapalapa, and next at Acatzintitlan, afterwards at Iztacal, and finally at the place where the hermitage of San Anton now is, at the entry of Mexico, and to that quarter which they now call San Pablo. During which time their idoll did comfort them in their travells and incoraged them, promising great matters.

CHAP. VII.—*Of the foundation of Mexico.*

The time being now come, that the father of lies should accomplish his promise made to his people, who could no longer suffer so many turnings, travells, and dangers, it happened that some old priests or sorcerers, being entred into a place full of water-lilies, they met with a very faire and cleere current of water, which seemed to be silver, and looking about, they found the trees, medowes, fish, and all that they beheld to be very white: wondring heereat, they remembred a prophecie of their god, whereby he had given them that for a token of their place of rest, and to make them Lords of other Nations. Then weeping for ioy, they returned to the people with these good newes. The night following, Vitzlipuztli appeared in a dreame to an antient priest, saying, that they should seeke out a Tunal[1] in the lake, which grew out of a stone (which as he told them, was the same place where by his commaundement they had cast the heart of Copil, sonne to the sorceresse, their enemy) and vpon this Tunal they should see a goodly Eagle, which fed on certaine small birdes. When they should see this, they should beleeve it was the place where their Cittie should be built, the which should surmount al others, and be famous throughout the world. Morning being come, the old man assembled the whole people, from the greatest to the least, making a long speech vnto them, how much they were bound vnto their god, and of the Revelation, which (although vnworthy) hee had received that night, concluding that all must seeke out that happie place which was promised them; which bred such devotion and ioy in them all, that presently they vndertooke the enterprise, and dividing themselves into bandes, they beganne to search, following the signes of the revelation of the desired place. Amidest the thickest of these water-lillies in the lake, they met with the same course of water they had seene the day before, but much differing, being not white, but red, like blood, the which divided it selfe into two streames, whereof the one was of a very obscure azure, the which bred admiration in them, noting some great mistery as they said. After much search heere and there, the Tunal appeared growing on a stone, whereon was a royall Eagle, with the wings displaied towardes the Sunne, receiving his heat. About this Eagle were many rich fethers, white, red, yellow, blew, and greene, of the same sort as they make their images, which Eagle held in his tallants a goodly birde. Those which sawe it and knew it to be the place fore-told by the Oracle, fel on their knees, doing great worship to the Eagle, which bowed the head looking on every side. Then was there great cries, demonstrations, and thanks vnto the Cre-

[1] Prickly pear.

ator, and to their great god Vitzlipuztli, who was their father, and had alwaies told them truth. For this reason they called the cittie which they founded there, Tenoxtitlan, which signifies Tunal on a stone, and to this day they carry in their armes, an Eagle upon a Tunal, with a bird in one tallant, and standing with the other vpon the Tunal. The day following, by common consent, they made an hermitage adioyning to the Tunal of the Eagle, that the Arke of their god might rest there, till they might have meanes to build him a sumptuous Temple : and so they made this hermitage of flagges and turfes covered with straw; then having consulted with their god, they resolved to buy of their neighbours, stone, timber, lime, in exchange of fish, frogges, and yong kids, and for duckes, water-hennes, curlews, and divers other kindes of sea fowles. All which things they did fish and hunt for in this Lake, whereof there is great aboundance. They went with these things to the markets of the Townes and Citties of the Tepanecas, and of them of Tezcuco their neighbours, and with pollicie they gathered together, by little and little, what was necessary for the building of their Cittie; so as they built a better Chappell for their idoll of lime and stone, and laboured to fill vp a great part of the lake with rubbish. This done, the idoll spake one night to one of his priests in these tearmes, " Say vnto the Mexicaines, that the Noblemen divide themselves everie one with their kinsfolkes and friends, and that they divide themselves into foure principall quarters, about the house which you have built for my rest, and let every quarter build in his quarter at his pleasure." The which was put in execution : and those be the foure principall quarters of Mexico, which are called at this day San Juan, Santa Maria la Redonda, San Pablo, and San Sebastian. After this, the Mexicaines being thus divided into these foure quarters, their god commanded them to divide amongest them the gods he should name to them, and that each principal quarter should name other special quarters, where these gods should be worshipped. So as vnder every one of these foure principall quarters, there were many less comprehended, according to the number of the idolls which their god commanded them to worship, which they called Calpultetco, which is as much as to say, god of the quarters. In this manner, the Citie of Mexico Tenoxtiltan was founded, and grew great.

Chap. VIII.—*Of the sedition of those of Tlatelulco, and of the first Kings the Mexicaines did choose.*

This division being made as afore-said, some olde men and Antients held opinion, that in the division, they had not respected them as they deserved: for this cause, they and their kinsfolke did mutine, and went to seeke another residence; and as they went thorough the lake, they found a small peece of ground or terrasse, which they call Tloteloli, where they inhabited, calling it Tlatellulco, which signifies place of a terrasse. This was the third division of the Mexicaines, since they left their Country. That of Mechoacan being the first, and that of Malinalco the second. Those which separated themselves and went to Tlatellulco were famous men, but of bad disposition; and therefore they practised against the Mexicaines, their neighbours, all the ill neighbourhood they could. They had alwaies quarrells against them, and to this day continues their hatred and olde leagues. They of Tenoxtiltan, seeing them of Tlatellulco thus opposite vnto them, and that they multiplied, feared that in time they might surmount them: heerevpon they assembled in counsell, where they thought it good to choose a King, whome they should obey, and strike terror into their enemies, that by this meanes they should bee more vnited and stronger among themselves, and their enemies not presume too much against them. Being thus resolved to choose a King, they took another advise very profitable and assured, to choose none among themselves, for the avoyding of dissentions, and to gaine (by their new King) some other neighbour nations, by whom they were invironed, being destitute of all succours. All well considered, both to pacifie the King of Culhuacan, whome they had greatly offended, having slaine and flead the daughter of his predecessor, and done him so great a scorne, as also to have a King of the Mexicaine blood, of which generation there were many in the Culhuacan, which continued there since the time they lived in peace amongst them; they resolved to choose for their King, a yong man called Acamapixtli, sonne to a great Mexicaine Prince, and of a Ladie, daughter to the King of Culhuacan. Presently they sent Ambassadors with a great present to demand this man, who delivered their Ambassage in these tearmes: "Great Lord, we your vassals and servants, placed and shut vp in the weedes and reedes of the Lake, alone and abandoned of all the Nations of the world, led onely and guided by our god to the place where we are, which falles in the iurisdiction of your limits of Ascapusalco, and of Tezcuco. Although you have suffered vs to live and remaine there, yet will we not, neither is it reason to, live without a head and lord to command, to correct, and governe vs, instructing vs in the course of our life, and defending vs from our enemies: Therefore we come to you, knowing that in your Court and house, there are children of our generation, linckt and alied with yours, issued from our entrailes, and yours, of our blood and yours, among the which we have knowledge of a grand-child of yours and ours, called Acamapixtli. We beseech you, therefore, to give him vs for Lord, we will esteeme him as hee deserves, seeing hee is of the lineage of the Lords of Mexico, and the Kings of Culhuacan.

The king having consulted vppon this poynt, and finding

it nothing inconvenient to be alied to the Mexicaines, who were valiant men, made them answer that they should take his grandchilde in good time, adding therevnto, that if he had beene a woman, hee woulde not have given her, noting the foule fact before spoken of, ending his discourse with these wordes: "Let my grand-childe go to serue your god, and be his lievetenant, to rule and governe his creatures, by whom we live, who is the Lord of night, day, and windes: Let him goe and be Lord of the water and land, and possesse the Mexicaine Nation, take him in good time, and vse him as my sonne and grand-childe." The Mexicaines gave him thanks, all ioyntly desiring him to marry him with his owne hand, so as he gave him to wife one of the noblest Ladies amongst them. They conducted the new King and Queene with all honour possible, and made him a solemne reception, going all in generall foorth to see the king, whom they led into pallaces, which were then but meane; and having seated them in royall throanes, presently one of the Antients and an Orator much esteemed amongest them, did rise vp, speaking in this manner: "My sonne, our Lord and King, thou art welcome to this poor house and citty, amongest these weedes and mudde, where thy poore fathers, grandfathers, and kinsfolkes, endure what it pleaseth the Lord of things created. Remember, Lord, thou commest hither to be the defence and support of the Mexicaine Nation, and to be the resemblance of our God Vitzlipuztli, wherevpon the charge and governement is given thee. Thou knowest we are not in our country, seeing the land we possesse at this day is anothers, neither know we what shall become of vs to-morrowe, or another day: Consider, therefore, that thou commest not to rest or recreate thy selfe, but rather to indure a new charge vnder so heavie a burden: wherein thou must continually labour, being slave to this multitude, which is fallen to thy lotte, and to all this neighbour people, whome they must strive to gratifie, and give them contentment, seeing thou knowest we live vpon their lands, and within their limites. And ending, hee repeated these wordes: "Thou art welcome, thou and the Queene our Mistris, to this your realme." This was the speech of the old man, which, with other orations (which the Mexicaine histories do celebrate) the children did vse to learne by hart, and so they were kept by tradition, some of them deserve well to be reported in their proper termes. The king aunswering, thanked them, and offered them his care and diligence in their defence and aide in all he could. After they gave him the othe, and after their maner set the royall crown vpon his head, the which is like to the Crowne of the dukes of Venice: the name of Acamapixtli, their first king, signifies a handfull of reeds, and therefore they carry in their armories a hand holding many arrows of reedes.

CHAP. IX.—*Of the strange tribute the Mexicaines paied to them of Azcapuzalco.*

The Mexicaines happened so well in the election of their new king, that in short time they grew to have some form of a common-weale, and to be famous among strangers; wherevpon their neighbours, moved with feare, practised to subdue them, especially the Tepanecas, who had Azcapuzalco for their metropolitane citty, to whome the Mexicaines payed tribute, as strangers dwelling in their land. For the king of Azcapuzalco fearing their power which increased, soght to oppresse the Mexicaines, and having consulted with his subjects, he sent to tel king Acamapixtli that the ordinary tribute they payed was too little, and that from thencefoorth they should bring firre trees, sapines, and willowes for the building of the citty, and moreover they shoulde make him a garden in the water planted with diverse kindes of hearbes and pulses, which they should bring vnto him yearely by water, dressed in this maner, without failing; which if they did not, he declared them his enemies, and would roote them out. The Mexicaines were much troubled at this commaundement, holding it impossible: and that this demaund was to no other end, but to seeke occasion to ruine them. But their god Vitzlipuztli comforted them, appearing that night to an olde man, commaunding him to say to the king his sonne in his name, that hee should make no difficultie to accept of this tribute, he would help them and make the meanes easie, which after happened: for the time of tribute being come, the Mexicanes carried the trees that were required, and moreover, a garden made and floating in the water, and in it much Mays (which is their corne) already grained and in the eare: there was also Indian pepper, beetes, Tomates, pease, gourds, and many other things, al ripe, and in their season. Such as have not seene the gardines in the lake of Mexico, in the middest of the water, will not beleeve it, but will say it is an inchantment of the Divell whom they worship: But in trueth it is a matter to be done, and there hath beene often seene of these gardens floating in the water; for they cast earth vpon reedes and grasse, in such sort as it never wastes in the water; they sowe and plant this ground, so as the graine growes and ripens very well, and then 'they remove it from place to place. But it is true, that to make this great garden easily, and to have the fruites grow well, is a thing that makes men iudge there was the worke of Vitzlipuztli, whom otherwise they call Patillas, specially having never made nor seene the like. The king of Azcapuzalco wondred much when he sawe that accomplished which he held impossible, saying vnto his subiects, that this people had a great god that made all easie vnto them, and hee sayd vnto the Mexicaines, that seeing their God gave them all things perfit hee would the yeare following, at the time of tribute, they shoulde bring in their gardine a wild ducke, and a heron, sitting on their egges, in such sorte, that they should hatch their yoong ones as they should arrive, without failing of a minute, vpon paine of his indignation. The Mexicans were much troubled and heavy with this prowde and strict commaunde: but their god, as he was accustomed, comforted them in the night, by one of his priests, saying that he would take all that charge vpon him, willing them not to fear, but beleeve that the day would come, whenas the Azcapuzalcos should pay with their lives this desire of new tributes. The time being come, as the Mexicaines carried all that was demaunded of their gardins, among the reeds and weeds of the gardin, they found a ducke and a heron hatching their egges, and at the same instant when they arrived at Azcapuzalco their yong ones were disclosed. Wherat the king of Azcapuzalco wondring beyond measure, he said againe to his people, that these were more than humane beings, and that the Mexicans beganne as if they would make themselves lordes over all those provinces. Yet did he not diminish the order of this tribute, and the Mexicans finding not themselves mighty enough, endured this subiection and slavery the space of fifty yeeres. In this time the king Acamapixtli died, having beautified the Citty of Mexico with many goodly buildings, streets, conduits of water, and great aboundance of munition. Hee raigned in peace and rest forty yeares, having bin always zealous for the good and increase of the common-weale.

As hee drew neare his end, hee did one memorable thing, that having lawfull children to whom he might leave the succession of the realme, yet would he not do it, but contrariwise hee spake freely to the common-weale, that as they had made a free election of him, so they should choose him that should seeme fittest for their good government, advising them therein to have a care to the good of the common-weale, and seeming grieved that he left them not freed from tribute and subiection, hee died, having recommended his wife and children vnto them, he left all his people sorrowfull for his death.

CHAP. X.—*Of the second King, and what happened in his raigne.*

The obsequies of the dead king performed, the Antients, the chiefe of the realme, and some part of the people assembled together to choose a King, where the Antients propounded the necessitie wherein they were, and that it was needefull to choose for chiefe of their citty, a man that had pity of age, of widows, and orphans, and to be a father of the commonweale: for in very deede they should be the feathers of his wings, the eie-browes of his eyes, and the beard of his face, that it was necessarie he were valiant, being needefull shortly to vse their forces, as their god had prophesied. Their resolution in the end was to chuse a sonne of the predecessor, vsing the like good office in accepting his sonne for successor, as hee had done to the commonweale, relying thereon. This young man was called Vitzilovitli, which signifieth a rich feather; they set the royall crowne vpon his head, and annointed him, as they have beene accustomed to doe to all their Kings, with an ointment they call Divine, being the same vnction wherewith they did annoynt their Idoll. Presently an Orator made an eloquent speech, exhorting him to arme himselfe with courage, and free them from the travells, slavery, and misery they suffered, being oppressed by the Azcapuzalcos: which done, all did him homage. This king was not married, and his Counsell helde opinion, that it was good to marry him with the daughter of the king of Azcapuzalco, to have him a friend by this alliance, and to obtain some diminution of their heavy burthen of tributes imposed vpon them, and yet they feared lest he should disdaine to give them his daughter, by reason they were his vassalls: yet the king of Azcapuzalco yeelded therevnto, having humbly required him, who, with curteous wordes, gave them his daughter, called Ayauchigual, whom they ledde with great pompe and ioy to Mexico, and performed the ceremony and solemnity of marriage, which was to tie a corner of the mans cloke to a part of the womans vaile in signe of the band of marriage. This Queene broght foorth a sonne, of whose name they demaunded advise of the king of Azcapuzalco, and casting lots as they had accustomed (being greatly given to soothsayings, especially vpon the names of their children), he would have his grand-childe called Chimalpopoca, which signifies a target casting smoke. The Queene, his daughter, seeing the contentment the King of Azcapuzalco had of his grand-child, tooke occasion to intreat him to releeve the Mexicaines of the heavy burthen of their tributes, seeing he had now a grand-child Mexicaine, the which the King willingly yeelded vnto, by the advise of his Counsell, granting (for the tribute which they paid) to bring yeerely a couple of duckes and some fish, in signe of subiection, and that they dwelt in his land. The Mexicaines, by this meanes, remained much eased and content, but it lasted little. For the Queene, their Protectrix, died soone after: and the yere following, likewise Vitzilovitli, the king of Mexico died, leaving his sonne, Chimalpopeca, tenne yeares olde; hee raigned thirteene yeeres, and died thirty yeeres old, or little more. Hee was held for a good king, and carefull in the service of his gods, whose Images hee held kings to be; and that the honour done to their god was done to the King who was his image. For this cause the kings have beene so affectionate to the service of their gods. This king was carefull to winne the love of his neighbours, and to trafficke with them, whereby hee augmented his citty, exercising his men in warrelike actions in the Lake, disposing them to that which he pretended, as you shall see presently.

CHAP. XI.—*Of Chimalpopoca, the third king, and his cruell death, and the occasion of warre which the Mexicaines made.*

The Mexicaines, for successor to their deceased king, did choose his sonne Chimalpopoca by common consent, although he were a child of tenne yeeres old, being of opinion that it was alwayes necessary to keepe the favor of the king of Azcapuzalco, making his grand-childe king. They then set him in his throane, giving him the ensignes of warre, with a bowe and arrowes in one hand, and a sword with rasours (which they commonly vse) in the right, signifying thereby (as they do say) that they pretended by armes to set themselves at liberty. The Mexicaines had great want of water, that of the Lake being very thicke and muddy, and therefore ill to drincke, so as they caused their infant king to desire of his grandfather, the king of Azcapuzalco, the water of the mountaine of Chapultepec, which is from Mexico a league, as is saide before, which they easely obtained, and by their industry made an aqueduct of faggots, weeds, and flagges, by the which they brought water to their citty. But because the Cittie was built within the Lake, and the aqueduct did crosse it, it did breake forth in many places, so as they could not inioy the water as they desired, and had great scarcitie: whervpon, whether they did expresly seeke it, to quarrell with the Tepanecas, or that they were mooved vppon small occasion, in the end they sent a resolute ambassage to the king of Azcapuzalco, saying they could not vse the water which he had gratiously granted them, and therefore they required him to provide them wood, lime, and stone, and to send his workmen, that by their meanes they might make a pipe of stone and lime that should not breake. This message nothing pleased the king, and much lesse his subiects, seeming to be too presumptuous a message, and purposely insolent, for vassals to their Lord. The chiefe of the Counsell disdaining thereat, said it was too bold that, not content with permission to live in an others land, and to have water given them, but they would have them goe to serve them: what a matter was that? And whereon presumed this fugitive nation, shut vp in the mud? They would let them know how fit they were to worke, and to abate their pride in taking from them their land and their lives.

In these termes and choller they left the king, whom they did somwhat suspect, by reason of his grandchild, and consulted againe anew what they were to doe, where they resolved to make a generall proclamation that no Tepaneca should have any commerce or trafficke with any Mexicaine, that they should not goe to their Cittie, nor receive any into theirs, vpon paine of death. Whereby we may vnderstand that the king did not absolutely commaund over his people, and that he governed more like a Consul or a Duke than a King, although since with their power the commaund of Kings increased, growing absolute Tyrants, as you shal see in the last Kings. For it hath beene an ordinarie thing among the Barbarians, that such as their power hath beene, such hath beene their commaund; yea, in our Histories of Spaine we finde in some antient kings that manner of rule which the Tepanecas vsed. Such were the first kings of the Romans, but that Rome declined from Kings to Consuls, and a Senate, till that after they came to be commaunded by Emperours. But these Barbarians, of temperate Kings became tyrants, of which governements a moderate monarchy is the best and most assured. But returne we now vnto our historie.

The king of Azcapuzalco seeing the resolution of his subiects, which was to kil the Mexicans, intreated them first to steale away the yong king, his grand-childe, and afterwards do what they pleased to the Mexicans. All in a manner yeelded heerevnto to give the king contentment, and for pitty they had of the child; but two of the chiefest were much opposite, inferring that it was bad counsell, for that Chimalpopoca, although hee were of their bloud, yet was it but by the mothers side, and that the fathers was to be preferred, and therefore they concluded that the first they must kill was Chimalpopoca, king of Mexico, protesting so to doe. The king of Azcapuzalco was so troubled with this contradiction, and the resolution they had taken, that soone after for very griefe he fell sicke and died. By whose death the Tepanecas, finishing their consultation, committed a notable treason; for one night the young king of Mexico sleeping without guard or feare of any thing, they of Azcapuzalco entred his pallace, and slew him sodainly, returning vnseene. The morning being come, when the Nobles went to salute the King, as they were accustomed, they found him slaine with great and cruell wounds; then they cried out, and filled all their cittie with teares: and transported with choller, they presently fell to armes, with an intent to revenge their Kings death. As they ranne vppe and downe, full of fury and disorder, one of their chiefest knightes stept foorth, labouring to appease them, with a grave admonition: "Whither goe you," saide hee, "O yee Mexicaines; quiet your selves, consider that things done without consideration are not well guided, nor come to good end: suppresse your griefe, considering that, although your king be dead, the noble blood of the Mexicaines is not extinct in him. Wee have children of our kings deceased, by whose conduct, succeeding to the realme, you shall the better execute what you pretend, having a leader to guide your enterprise, go not blindely, surcease, and choose a king first to guide and encourage you against your enemies. In the meane time dissemble discreetly, performing the funeralls of your deceased king, whose body you see heere present, for heereafter you shall finde better meanes to take revenge." By this meanes, the Mexicaines passed no farther, but stayed to make the obsequies of their King, wherevnto they invited the Lords of Tezcuco and Culhuacan, reporting vnto them this foule and cruell fact, which the Tepanecas had committed, moving them to have pitty on them, and incensing them against their enemies, concluding that their resolution was to die or to bee revenged of so great an indignitie, intreating them not to favour so vniust a fact of their enemies; and that for their part, they desired not their aide of armes or men, but onely to bee lookers on of what should passe, and that for their maintenance they would not stoppe nor hinder the commerce as the Tepanecas had done. At these speeches they of Tezcuco and Culhuacan made them great shewes of good will, and that they were well satisfied, offering them their citties, and all the commerce they desired, that they might provide vittaile and munition at their pleasure, both by land and water. After this, the Mexicanes intreated them to stay with them, and assist at the election of their King; the which they likewise granted, to give them contentment.

CHAP. XII.—*Of the fourth King, called Iscoalt, and of the warre against the Tepanecas.*

The Electors being assembled, an old man that was held for a great Orator, rose vp, who, as the histories report, spake in this manner: "The light of your eyes, O Mexicaines, is darkened, but not of your hearts: for although you have lost him that was the light and guide of the Mexicaine Common-weale, yet that of the heart remaines: to consider, that although they have slaine one man, yet there are others that may supply with advantage the want we have of him: the Mexicaine Nobilitie is not extinguished thereby, nor the blood royall decaied. Turne your eyes and looke about you;

you shall see the Nobilitie of Mexico set in order, not one nor two, but many and excellent Princes, sonnes to Acamapixtli, our true and lawfull King and Lord. Heere you may choose at your pleasure, saying, I will this man, and not that. If you have lost a father, heere you may find both father and mother: make account, O Mexicaines, that the Sunne is eclipsed and darkened for a time, and will returne suddenly. If Mexico hath beene darkened by the death of your King, the Sunne will soon shew, in choosing another King. Looke to whom, and vpon whom you shall cast your eyes, and towards whom your heart is inclined, and this is hee whom your god Vitzlipuztli hath chosen." And continuing a while this discourse, he ended to the satisfaction of all men. In the end, by the consent of this Counsell, Izcoalt was chosen King, which signifies a snake of rasors,[1] who was sonne to the first King Acamapixtli, by a slave of his : and although he were not legitimate, yet they made choyce of him, for that he exceeded the rest in behaviour, valour, and magnanimitie of courage. All seemed very well satisfied, and above all, these of Tescuco, for their king was married to a sister of Iscoalt. After the King had beene crowned and set in his royall seat, another Orator stept up, discoursing how the king was bound to his Common-weale, and of the courage he ought to shew in travell, speaking thus: "Behold this day we depend on thee; it may be thou wilt let fall the burthen that lies vpon thy shoulders, and suffer the old man and woman, the orphan and the widowe to perish. Take pittie of the infants that go creeping in the ayre, who must perish if our enemies surmount vs; vnfold then and stretch forth thy cloake, my Lord, to beare these infants vpon thy shoulders, which be the poore and the common people, who live assured under the shadowe of thy wings, and of thy bountie." Vttering many other words vpon this subiect, the which (as I have

[1] "Culebra de navajas."

said) they learne by heart, for the exercise of their children, and after did teach them as a lesson to those that beganne to learne the facultie of Orators. In the meane time, the Tepanecas were resolute to destroy the Mexicaines, and to this end they had made great preparations. And therefore the new King tooke counsell for the proclaiming of warre, and to fight with those that had so much wronged them. But the common people, seeing their adversaries to exceede them farre in numbers and munition for the warre, they came amazed to their King, pressing him not to vndertake so dangerous a warre, which would destroy their poor Cittie and Nation: wherevpon being demaunded what advise were fittest to take, they made answer that the King of Azcapuzalco was very pittifull, that they should demand peace, and offer to serve him, drawing them forth those marshes, and that he should give them houses and lands among his subiects, that by this meanes they might depend all vppon one Lord. And for the obtaining heereof, they should carry their god in his litter for an intercessor. The cries of the people were of such force (having some Nobles that approved their opinion), as presently they called for the Priests, preparing the litter and their god, to perform the voyage. As this was preparing, and every one yeelded to this treatie of peace, and to subiect themselves to the Tepanecas, a gallant yong man, and of good sort, stept out among the people, who, with a resolute countenance, spake thus vnto them: "What meanes this, O yee Mexicaines, are yee mad? How hath so great cowardise crept in among vs? Shall we go and yeeld ourselves thus to the Azcapuzalcos." Then turning to the King, he said: "How now, my Lord, will you endure this? Speak to the people, that they may suffer vs to finde out some meanes for our honour and defence, and not to yeelde our selves so simply and shamefully into the hands of our enemies." This yong man was called Tlacaellel, nephew to the King, he was the most valiant Captaine and greatest Counsellor that ever the Mexicaines had, as you shall see heereafter. Izcoalt, incouraged by that his nephew had so wisely spoken, retained the people, saying they should first suffer him to try another better meanes. Then turning towards his Nobilitie, he said vnto them : " You are all heere, my kinsmen, and the best of Mexico, hee that hath the courage to carrie a message to the Tepanecas, let him rise vp." They looked one vpon another, but no man stirred nor offered himselfe to the word. Then this yong man, Tlacaellel, rising, offered himselfe to go, saying, that seeing he must die, it did import little whether it were to-day or to-morrow : for what reason should he so carefully preserve himselfe ? he was therefore readie, let him command what he pleased. And although all held this for a rash attempt, yet the King resolved to send him, that he might thereon vnderstand the will and disposition of the King of Azcapuzalco and of his people; holding it better to hasten his nephew's death, then to hazard the honour of his Common-weale. Tlacaellel being ready, tooke his way, and being come to the gards, who had commaundement to kill any Mexicaines that came towards them by cunning or otherwise : he perswaded them to suffer him to passe to the king, who wondered to see him, and hearing his ambassage, which was to demand peace of him vnder honest conditions, answered, that hee would impart it to his subiects, willing him to returne the next day for his answer ; then Tlacaellel demanded a passport, yet could he not obtaine any, but that he should vse his best skill. With this he returned to Mexico, giving his words to the guards to returne. And, although the King of Azcapuzalco desired peace, being of a milde disposition, yet his subiects did so incense him, as his answer was open warre. The which being heard by the messenger, he did all his King commanded him, declaring by this ceremony to give armes, and anointing the King with the vnction of the dead, that in his Kings behalfe he did defie him. Having ended all, the King of Azcapuzalco suffering himselfe to be anointed and crowned with feathers, giving goodly armes in recompence to the messenger, wishing him not to returne by the pallace gate, whereas many attended to cut him in peeces, but to go out secretly by a little false posterne that was open in one of the courts of the Pallace. This yong man did so, and turning by secret waies, got away in safetie in sight of the guards, and there defied them, saying, "Tepanecas and Azcapuzalcas, you do your office ill; vnderstand you shall all die, and not one Tepaneca shall remaine alive." In the meane time the guardes fell vpon him, where he behaved him selfe so valiantly, that hee slew some of them: and seeing many more of them come running, hee retyred himselfe gallantly to the Cittie, where he brought newes that warre was proclaimed with the Tepanecas, and that hee had defied their King.

CHAP. XIII.—*Of the battell the Mexicaines gave to the Tepanecas, and of the victorie they obtained.*

The defie being knowne to the Commons of Mexico, they came to the king, according to their accustomed cowardise, demaunding leave to departe the Citty, holding their ruin certaine. The king didde comfort and incourage them, promising to give them libertie if they vanquished their enemies, willing them not to feare. The people replied: "And if we be vanquished what shall we doe?" "If we be overcome (aunswered the king) we will be bound presently to yeeld ourselves into your hands to suffer death, eate our flesh in your dishes, and be revenged of vs." "It shall be so then (saide they) if you loose the victorie, and if you obtain the victorie, we do presently offer our selves to be your Tributaries, to labour in your houses, to sowe your ground, to carrie your armes and baggage when you goe to the warres for ever, wee and our descendants after vs." These accordes made betwixt the people and the nobilitie (which they did after fully performe, eyther willingly or by constraint, as they had promised), the king named for his captain generall Tlacaellel, the whole camp was put in order, and into squadrons, giving the places of captaines to the most valiant of his kinsfolkes and friends: then did hee make them a goodly speech, whereby he did greatly incourage them, being now wel prepared, charging all men to obey the commaundement of the Generall whome he had appoynted: he divided his men into two partes, commanding the most valiant and hardie to give the first charge with him, and that all the rest should remaine with the king Izcoatl, vntil they should see the first assaile their enemies. Marching then in order, they were discovered by them of Azcapuzalco, who presently came furiously foorth the citty, carrying great riches of gold, silver, and armes of great value, as those which had the empire of all that country. Izcoatl gave the signall to battaile, with a little drumme he carried on his shoulders, and presently they raised a general showt, crying Mexico, Mexico, they charged the Tapanecans, and although they were farre more in number, yet did they defeate them, and force them to retire into their Cittie; then advaunced they which remained behinde, crying Tlacaellel, victorie, victorie, all sodainely entred the Citty, where (by the Kings commandement) they pardoned not any man, no not olde men, women, nor children, for they slew them all, and spoyled the Citty, being very rich. And not content heerewith, they followed them that fled, and were retired into the craggy rocks of the Sierras or neere mountaines, striking and making a great slaughter of them. The Tapanecans being retired to a mountaine, cast downe their armes, demaunding their lives, and offering to serve the Mexicaines, to give them lands and gardins, stone, lime and timber, and to hold them

I I

alwayes for their Lordes. Vpon this condition Tlacaellec retired his men, and ceased the battell, graunting them their lives upon the former conditions, which they did solemnely sweare. Then they returned to Azcapuzalco, and so with their rich and victorious spoiles to the cittie of Mexico. The day following the king assembled the Nobilitie and the people, to whom he laid open the accord the Commons had made, demaunding of them if they were content to persist therin: the Commons made answer that they had promised, and they had well deserved it, and therefore they were content to serve them perpetually. Wherevpon they took an othe, which since they have kept without contradiction.

This done, Izcoatl returned to Azcapuzalco (by the advise of his counsell), he divided all the lands and goods of the conquered among the conquerours, the chiefest parte fell to the King, then to Tlacaellel, and after to the rest of the Nobles, as they best deserved in the battell. They also gave land to some plebeians, having behaved themselves valiantly; to others they distributed the pillage, making small account of them as of cowardes. They appointed lands in common for the quarters of Mexico, to every one his part, for the service and sacrifices of their gods. This was the order, which after they alwayes kept, in the division of the lands and spoyles of those they had vanquished and subdewed. By this meanes they of Azcapuzalco remained so poore, as they had no lands left them to labor, and (which was worse) they tooke their king from them, and all power to chuse any other then him of Mexico.

Chap. xiv.—*Of the warre and victory the Mexicaines had against the Cittie of Cuyoacan.*

Although the chiefe cittie of the Tepanecas was that of Azcapuzalco, yet had they others with their private Lordes, as Tucuba and Cuyoacan. These seeing the storme passed, would gladly that they of Azcapuzalco had renewed the warre against the Mexicans, and seeing them danted, as a nation wholy broken and defeated, they of Cuyoacan resolved to make warre by themselves; to the which they laboured to draw the other neighbor nations, who would not stir nor quarrell with the Mexicans. In the meane time the hatred and malice increasing, they of Cuyoacan beganne to ill intreate the women that went to their markets, mocking at them, and doing the like to the men over whom they had power: for which cause the king of Mexico defended,[1] that none of his should goe to Cuyoacan, and that they should receive none of them into Mexico, the which made them of Cuyoacan resolve wholy to warre: but first they would provoke them by some shamefull scorne, which was, that having invited them to one of their solemn feasts, after they had made them a goodly banquet, and feasted them with a great daunce after their manner, they sent them, for their fruite, womens apparell, forcing them to put it on, and so to returne home like women to their cittie, reproching them, that they were cowards and effeminate, and that they durst not take armes, being sufficiently provoked. Those of Mexico say, that for revenge they did vnto them a fowle scorne, laying at the gates of their cittie of Cuyoacan certaine things which smoaked,[2] by meanes whereof many women were delivered before their time, and many fell sicke. In the end, all came to open warre, and there was a battell fought, wherein they imployed all their forces, in

[1] "Vedò." [2] "Ciertos humazos."

the which Tlacaellel, by his courage and policie in warre, obtained the victory. For, having left king Izcoatl in fight with them of Cuyoacan, he put himselfe in ambush with some of the most valiant souldiers, and so turning about charged them behind, and forced them to retire into their citty. But seeing their intent was to flie into a temple, which was verie strong, he, with three other valiant souldiers, pursued them eagerly, and got before them, seising on the temple and firing it, so as he forced them to flie to the fields, where he made a great slaughter of the vanquished, pursuing them two leagues into the countrey, vnto a litle hill, where the vanquished, casting away their weapons and their armes across, yeelded to the Mexicans, and with many teares craved pardon of their overweening follie, in vsing them like women, offering to bee their slaves: so as, in the end, the Mexicaines did pardon them. Of this victory the Mexicaines carried away very rich spoiles of garments, armes, gold, silver, iewells, and rich feathers, with a great number of captives. In this battaile there were three of the principals of Culhuacan that came to aide the Mexicaines to winne honour, the which were remarkable above all. And since being knowen to Tlacaellel, and having made proofe of their fidelitie, he gave them Mexicaine devises, and had them alwayes by his side, where they fought in all places very valiantly. It was apparant that the whole victory was due to the Generall and to these three; for, among so many captives taken, two third partes were wonne by these foure, which was easily knowen by a policie they vsed: for, taking a captive, they presently cut off a little of his haire and gave it to others, so as it appeared that those which had their haire cut, amounted to that number, whereby they wonne great reputation and fame of valiant men. They were honoured as conquerors, giving them good portions of the spoiles and lands, as the Mexicans have alwayes vsed to doe, which gave occasion to those that did fight to become famous, and to winne reputation by armes.

Chap. xv.—*Of the warre and victorie which the Mexicans had against the Suchimilcos.*

The Nation of the Tepanecas being subdewed, the Mexicaines had occasion to do the like to the Suchimilcos, who (as it hath beene saide) were the first of the seven caves or lineages that peopled this land. The Mexicans sought not the occasion, although they might presume as conquerors to extend their limits, but the Suchimilcos didde moove them, to their owne ruine, as it happens to men of small iudgement that have no foresight, who not preventing the the mischefe they imagined, fall into it. The Suchimilcos held opinion that the Mexicans, by reason of their victories past, should attempt to subdue them, and consulted heereon amongst themselves. Some among them thought it good to acknowledge them for superiors, and to applaude their good fortune, but the contrary was allowed, and they went out to give them battel; which Izcoatl the king of Mexico vnderstanding, he sent his Generall Tlacaellel against them, with his army; the battell was fought in the same field that divides their limites, which two armies were equall in men and armes, but very divers in their order and manner of fighting; for that the Suchimilcos charged all together on a heape confusedly, and Tlacaellel divided his men into squadrons with a goodly order, so as he presently brake his ennemies, forcing them to retire into their cittie, into the which they entred, following them to the Temple whither they fled, which they fiered, and forcing them to flie vnto the mountaines; in the end they brought them to this poynt, that they yeelded with their armes acrosse. The Generall Tlacaellel returning in great triumph, the priests went foorth to receive him, with their musicke of flutes, and giving incense. The chiefe Captaines vsed other ceremonies and shews of ioy, as they had bin accustomed to doe, and the king with all the troupe went to the Temple, to give thanks to their false god, for the divell hath alwayes beene very desirous hereof, to challenge to himselfe the honor which he deserves not, seeing it is the true God which giveth victories, and maketh them to rule whome he pleaseth. The day following king Izcoatl went vnto the citty of Suchimilco, causing himselfe to be sworne king of the Suchimilcos; and for their comfort he promised to doe them good. In token whereof hee commaunded them to make a great cawsey stretching from Mexico to Suchimilco, which is foure leagues, to the end there might bee more commerce and trafficke amongest them. Which the Suchimilcos performed, and in shorte time the Mexicaine governement seemed so good vnto them, as they helde themselves happy to have changed their king and commonweale. Some neighbors, pricked forward by envy or feare to their ruines, were not yet made wise by others miseries.

Cuitlavaca was a citty within the lake, which though the name and dwelling be chaunged, continueth yet. They were active to swimme in the lake, and therefore they thought they might much indomage and annoy the Mexicaines by water, which the King vnderstanding, hee resolved to send his army presently to fight against them. But Tlacaellel little esteeming this warre, holding it dishonorable to lead an army against them, made offer to conquer them with the children onely, which he performed in this maner; he went vnto the Temple and drew out of the Convent such children as he thought fittest for this action, from tenne to eighteene yeeres of age, who knew how to guide their boates or canoes, teaching them certaine pollicies. The order they held in this warre was, that he went to Cuitlavaca with his children, where by his pollicy hee pressed the ennemy in such sorte, that hee made them to flie; and as he followed them, the lord of Cuitlavaca mette him and yeelded vnto him, himselfe, his citty, and his people, and by this meanes he stayed the pursuite. The children returned with much spoyle, and many captives for their sacrifices, being solemnely received with a great procession, musike and perfumes, and they went to worshippe their gods, in taking of the earth which they did eate, and drawing blood from the forepart of their legges with the Priests lancets, with other superstitions which they were accustomed to vse in the like solemnities. The children were much honoured and incoraged, and the king imbraced and kissed them, and his kinsmen and alies accompanied them. The bruite of this victorie ranne throughout all the country, how that Tlacaellec had subdued the city of Cuitlavaca with children; the news and consideration whereof opened the eyes of those of Tezcuco, a chiefe and very cunning Nation for their manner of life; So as the king of Tezcuco was first of opinion, that they should subiect themselves to the king of Mexico, and invite him therevnto with his cittie. Therefore by the advise of his Counsell, they sent Ambassadors, good Orators, with honorable presents, to offer themselves vnto the Mexicans, as their subiects, desiring peace and amitie, which was gratiously accepted; but by the advise of Tlacaellec he vsed a ceremony for the effecting thereof, which was that those of Tezcuco should come forth armed against the Mexicans, where they should fight, and presently yeelde, which was an act and ceremony of warre, without any effusion of bloud on either side. Thus the king of Mexico became soveraigne Lord of Tezcuco, but hee tooke not their king from them, but made him of his privie counsell, so as they have alwayes maintained themselves in this manner vntill the time of Moteçuma the second, during whose raigne the Spaniards entred. Having subdued the land and citty of Tezcuco, Mexico remained Lady and Mistris of all the landes and citties about the Lake, where it is built. Izcoatl having enioyed this prosperitie, and raigned twelve yeeres, died, leaving the realme which had beene given him much augmented by the valour and counsell of his nephew Tlacaellel (as hath afore beene saide) who held it best to choose an other king then himselfe, as shall heereafter be shewed.

CHAP. XVI.—*Of the fift King of Mexico, called Monteçuma, the first of that name.*

Forasmuch as the election of the new King belonged to foure chiefe Electors (as hath been said), and to the King of Tezcuco, and the King of Tacuba, by especiall priviledge; Tlacaellel assembled these six personages, as he that had the soveraigne authoritie, and having propounded the matter vnto them, they made choise of Monteçuma, the first of that name, nephew to the same Tlacaellel. His election was very pleasing to them all, by reason whereof they made most solemne feasts, and more stately then the former. Presently after his election, they conducted him to the Temple with a great traine, where before the divine harth (as they call it) where there is continuall fire, they set him in his royall throne, putting vpon him his royall ornaments. Being there, the King drew blood from his eares and the calves of his legs, and his shins, with certain pointed instruments of a tiger and of a deer, used for that purpose, which was the sacrifice wherein the divell delighted to be honoured. The Priests, Antients, and Captaines made their orations, all congratulating his election. They were accustomed in their elections to make great feasts and dances, where they wasted many lightes. In this Kings time the custome was brought in, that the King should go in person to make warre in some province, and bring captives to solemnize the feast of his coronation, and for the solemne sacrifices of that day. For this cause King Monteçuma went into the province of

Chalco, inhabited by a warlike people; from whence (having fought valiantly) he brought a great number of captives, whereof he did make a notable sacrifice the day of his coronation, although at that time he did not subdue all the province of Chalco, being a very warlike nation. Many came to this coronation from divers provinces, as well neere as farre off, to see the feast, at the which all commers were very bountifully entertained and clad, especially the poore, to whom they gave new garments. For this cause they brought that day into the cittie, the Kings tributes, with a goodly order, which consisted in stuffes to make garments of all sorts, in cacao, gold, silver, rich feathers, great burthens of cotton, cucumbers, sundry sortes of pulses, many kindes of sea fish, and of the fresh water, great store of fruites, and venison without number, not reckoning an infinite number of presents, which other kings and lords sent to the new king. All this tribute marched in order according to the provinces, and before them the stewards and receivers, with divers markes and ensignes, in very goodly order; so as it was one of the goodliest things of the feast, to see the entry of the tribute. The King being crowned, he imploied himselfe in the conquest of many provinces, and for that he was both valiant and vertuous, hee still increased more and more, vsing in all his affaires the counsell and industry of his generall Tlacaellel, whom he did alwaies love and esteeme very much, as hee had good reason. The warre wherein hee was most troubled and of greatest difficultie, was that of the province of Chalco, wherein there happened great matters, whereof one was very remarkable, which was, that they of Chalco had taken a brother of Monteçuma in the warres, whome they resolved to choose for their king, asking him very curteously if he would accept of this charge. He answered (after much importunity, still persisting therein), that if they meant plainely to choose him for their king, they should plant in the market place a tree or very high stake, on the toppe whereof they should make a little scaffold, and meanes to mount vnto it. The Chalcos supposing it had beene some ceremony to make himselfe more apparent, presently effected it; then assembling all his Mexicaines about the stake, he went to the toppe with a garland of flowers in his hand, speaking to his men in this maner, "O valiant Mexicaines, these men will choose mee for their King; but the gods will not permit that to be a King I should committe any treason against my countrie, but contrariwise, I wil that you learne by me that it behoveth vs rather to indure death then to ayde our enemies." Saying these wordes he cast himselfe downe, and was broken in a thousand peeces, at which spectacle the Chalcos had so great horror and dispite, that presently they fell vpon the Mexicaines and slew them all with their launces, as men whom they held too prowde and inexorable, saying, they had divelish hearts. It chanced the night following, they heard two owles making a mournefull cry, which they did interpret as an vnfortunate signe, and a presage of their neere destruction, as it succeeded; for King Monteçuma went against them in person with all his power, where he vanquished them, and ruined all their kingdome; and passing beyond the Sierra Nevada, hee conquered still even vnto the North sea. Then returning towards the South sea, hee subdued many provinces, so as he became a mighty King, all by the helpe and counsell of Tlacaellel, who in a manner conquered all the Mexicaine nation. Yet hee held an opinion (the which was confirmed) that it was not behoovefull to conquer the province of Tlascalla, that the Mexicaines might have a fronter enemy, to keepe the youth of Mexico in exercise and allarme; and that they might have numbers of captives to sacrifice to their idols, wherein they did waste (as hath beene said) infinite numbers of men, which should bee taken by force in the warres. The honour must be given to Monteçuma, or to speake truly, to Tlacaellel his Generall, for the good order and policy setled in the realme of Mexico, as also for the counsells and goodly enterprises which they did execute; and likewise for the numbers of Iudges and Magistrates, being as well ordered there as in any common-weale; yea, were it in the most flourishing of Europe. This King did also greatly increase the King's house, giving it great authoritie, and appointing many and sundry officers, which served him with great pompe and ceremony. Hee was no lesse remarkable touching the devotion and service of his idolls, increasing the number of his Ministers and instituting new ceremonies, wherevnto hee carried a great respect.

Hee built that great temple dedicated to their god Vitzilipuztli, whereof is spoken in the other Booke. He did sacrifice at the dedication of this temple, a great number of men, taken in sundry victories: finally inioying his Empire in great prosperitie; hee fell sicke, and died, having raigned twenty-eight yeares, vnlike to his successor Ticocic, who did not resemble him, neither in valour, nor in good fortune.

CHAP. XVII.—*How Tlacaellel refused to be King, and of the election and deedes of Ticocic.*

The foure Deputies assembled in counsell, with the lords of Tezcuco and Tacuba, where Tlacaellel was President in the election, where by all their voices Tlacaellel was chosen, as deserving this charge better than any other. Yet he refused it, perswading them by pertinent reasons that they should choose another, saying, that it was better and more expedient to have another king, and he to be his instrument and assistant, as hee had beene till then, and not to lay the whole burthen vpon him, for that he held himselfe no lesse bound for the Common-weale, then if hee were king. It is a rare thing to refuse principalitie and commaund, and to indure the paine and the care, and not to reape the honour. There are few that will yeeld vp the power and authority which they may hold, were it profitable to the common-weale. This Barbarian did heerein exceed the wisest amongst the Greekes and Romans, and it may be a lesson to Alexander and Iulius Cæsar, whereof the one held it little to command the whole world, putting his most deere and faithfull servants to death vpon some small iealosies of rule and empire: and the other declared himselfe enemy to his country, saying, that if it were lawfull to do anything against law and reason, it was for a kingdome: such is the thirst and desire of commaund. Although this acte of Tlacaellels might well proceede from too great a confidence of himselfe, seeming to him, though he were not king, yet in a maner that he commanded kings, suffering him to carry certaine markes, as a tiara or ornament for the head, which belonged onely to themselves. Yet this act deserves greater commendation, and to be well considered of, in that he held opinion to be better able to serve his common-weale as a subiect, then being a soveraigne Lord. And as in a comedie he deserves most commendation that represents the personage that imports most, bee it of a sheepheard or a peasant, and leave the King or Captaine to him that can performe it: so in good Philosophy, men ought to have a special regard to the common good, and apply themselves to that office and place which they best vnderstand. But this philosophie is farre from that which is practised at this day. But let vs return to our discourse, and say, that in recompense of his modestie, and for the respect which the Mexicaine Electors bare him, they demanded of Tlacaellel (that seeing he would not raigne) whom he thought most fit: wherevpon he gave his voice to a sonne of the deceased king, who was then very yong, called Ticocic: but they replied that his shoulders were very weake to beare so heavy a burthen. Tlacaellel answered that he was there to help him to beare the burthen, as he had done to the deceased: by meanes whereof they tooke their resolution, and Ticocic was chosen, to whom were done all the accustomed ceremonies.

They pierced his nosthrils, and for an ornament put an emerald therein: and for this reason, in the Mexicaine bookes, this king is noted by his nosthrills pierced. Hee differed much from his father and predecessor, being noted for a coward, and not valiant. He went to make warre for his coronation, in a province that was rebelled, where he lost more of his own men then hee tooke captives; yet he returned, saying, that hee brought the number of captives required for the sacrifice of his coronation, and so hee was crowned with great solemnitie. But the Mexicaines, discontented to have a king so little disposed to warre, practised to hasten his death by poison. For this cause hee continued not above foure yeares in the kingdome: whereby wee see that the children do not alwaies follow the blood and valour of their fathers; and the greater the glorie of the predecessors hath beene, the more odious is the weakenes and cowardise of such that succeed them in command, and not in merit. But this losse was well repaired by a brother of the deceased, who was also sonne to great Monteçuma, called Axayaca, who was likewise chosen by the advice of Tlacaellel, wherein hee happened better than before.

CHAP. XVIII.—*Of the death of Tlacaellel, and the deedes of Axayaca, the seventh King of Mexicaines.*

Now was Tlacaellel very old, who by reason of his age, he was carried in a chaire upon mens shoulders, to assist in counsell when busines required. In the end hee fell sicke, whenas the king (who was not yet crowned), did visit him often, sheading many teares, seeming to loose in him his father, and the father of his countrie. Tlacaellel did most affectionately recommend his children vnto him, especially the eldest, who had showed himselfe valiant in the former warres. The king promised to have regard vnto him, and the more to comfort the olde man, in his presence he gave him the charge and ensignes of Captaine Generall, with all the pre-eminences of his father; wherewith the old man remained so well satisfied, as with this content he ended his daies. If hee had not passed to another life, they might have held themselves very happy, seeing that of so poore and small a cittie, wherein he was borne, he established, by his valour and magnanimitie, so great, so rich, and so potent a king-dome. The Mexicans made his funerall, as the founder of that Empire, more sumptuous and stately, then they had done to their former kings. And presently after Axayaca, to appease the sorrow which all the people of Mexico shewed for the death of their captaine, resolved to make the expedition necessary for his coronation. Hee therefore led his army with great expedition into the province of Tehuantepec, two hundred leagues from Mexico, where he gave battaile to a mighty army and an infinite number of men assembled together, as well out of that province, as from their neighbours, to oppose themselves against the Mexicans. The first of his campe that advanced himselfe to the combate, was the King himselfe, defying his ennemies, from whome hee made shewe to fly when they charged him, vntill he had drawne them into an ambuscadoe, where many souldiers lay hidden vnder straw, who suddenly issued forth, and they which fled, turned head: so as they of Tehuantepec remained in the midst of them, whom they charged furiously, making a great slaughter of them: and following their victory, they razed their citty and temple, punishing all their neighbours rigorously. Then went they on farther, and without any stay, conquered to Guatulco,

WAR WITH TLATELLULCO.

the which is a port at this day well knowne in the South sea. Axayaca returned to Mexico with great and rich spoiles, where he was honourably crowned, with sumptuous and stately preparation of sacrifices, tributes, and other things, whither many came to see his coronation. The Kings of Mexico received the crowne from the hands of the King of Tezcuco, who had the preeminence. He made many other enterprises, where he obtained great victories, being alwaies the first to leade the army, and to charge the enemy; by the which hee purchased the name of a most valiant captaine: and not content to subdue strangers, he also suppressed his subiects which had rebelled, which never any of his predecessors ever could doe, or durst attempt. We have already shewed how some seditious of Mexico had divided themselves from that common-weale, and built a cittie neare vnto them, which they called Tlatellulco, whereas now Santiago is.

These being revolted, held a faction aparte, and encreased and multiplied much, refusing to acknowledge the kings of Mexico, nor to yeeld them obedience. The king Axayaca sent to advise them not to live divided, but being of one bloud, and one people, to ioyne together, and acknowledge the king of Mexico: wherevpon the Lorde of Tlatellulco made an aunswere full of pride and disdaine, defieing the king of Mexico to single combat with himselfe: and presently mustred his men, commaunding some of them to hide themselves in the weeds of the Lake; and the better to deceive the Mexicans, he commaunded them to take the shapes of ravens, geese, and other beasts, as frogs, and such like, supposing by this meanes to surprise the Mexicans as they should passe by the waies and cawsies of the Lake. Having knowledge of this defiance, and of his adversaries policie, he divided his army, giving a part to his generall, the sonne of Tlacaellel, commaunding him to charge this ambuscadoe in the Lake; and he with the rest of his people, by an vnfrequented way, went and incamped before Tlatellulco. Presently hee called him who had defied him to performe his promise, and as the two Lordes of Mexico and Tlatellulco advaunced, they commaunded their subiects not to moove, vntill they had seene who should be conquerour, which was done, and presently the two Lordes incountered valiantly, where having fought long, in the end the Lorde of Tlatellulco was forced to turne his backe, being vnable to indure the furious charge of the king of Mexico. Those of Tlatellulco seeing their captaine flie, fainted, and fled likewise, but the Mexicans following them at the heeles, charged them furiously: yet the Lord of Tlatellulco escaped not the hands of Axayaca, for thinking to save himselfe, he fled to the toppe of the temple, but Axayaca folowed him so neere, as he seised on him with great force, and threw him from the toppe to the bottome, and after set fire on the temple and the cittie. Whilest this passed at Tlatellulco, the Mexicane generall was very hote in the revenge of those that pretended to defeate him by pollicie, and after he had forced them to yeelde, and to cry for mercy, the general sayed he would not pardon them vntil they had first performed the offices of those figures they represented, and therefore he would have them crie like frogges and ravens, every one according to the figure which he had vndertaken, else they had no composition: which thing he did to mocke them with their own policie. Feare and necessitie be perfect teachers; so as they did sing and crie with all the differences of voyces that were commaunded them, to save their lives, although they were much grieved at the sport their enimies made at them. They say that vnto this day, the Mexicans vse to ieast at the Tlatellulcans, which they beare impatiently, when they putte them in minde of this singing and crying of beasts. King Axayaca tooke pleasure at this scorne and disgrace, and presently after they returned to Mexico with great ioy. This king was esteemed for one of the best that had commaunded in Mexico. Hee raigned eleaven yeares, and one succeeded that was much inferiour vnto him in valour and vertue.

Chap. xix.—*Of the deedes of Autzol the eighth King of Mexico.*

Among the foure Electors that had power to chuse whome they pleased to be king, there was one indued with many perfections, named Autzol. This man was chosen by the rest, and this election was very pleasing to all the people : for besides that he was valiant, all held him curteous and affable to every man, which is one of the chief qualities required in them that commaund, to purchase love and respect. To celebrate the feast of his coronation, hee resolved to make a voyage, and to punish the pride of those of Quaxutatlan, a very rich and plentifull province, and at this day the chiefe of New Spaine. They had robbed his officers and stewards, that carried the tribute to Mexico, and therewithall had rebelled. There was great difficulty to reduce this Nation to obedience, lying in such sort, as an arme of the sea stopt the Mexicans passage: to passe the which, Autzol (with a strange device and industry) caused an Iland to be made in the water, of faggots, earth, and other matter; by meanes whereof, both hee and his men might passe to the enemy, where giving them battell, he conquered them and punished them at his pleasure. Then returned hee vnto Mexico in triumph, and with great riches, to bee crowned King, according to their custome. Autzol extended the limits of his kingdome farre, by many conquests, even vnto Guatimala, which is three hundred leagues from Mexico. He was no less liberall than valiant: for whenas the tributes arrived (which as I have saide) came in great aboundaunce, hee went foorth of his pallace, gathering together all the people into one place, then commaunded he to bring all the tributes, which hee divided to those that had neede. To the poore he gave stuffes to make apparrell, and meate, and whatsoever they had neede of in great aboundaunce, and things of value, as golde, silver, iewels, and feathers, were divided amongst the captaines, souldiers, and servants of his house, according to every man's merite. This Autzol was likewise a great polititian, hee pulled downe the houses ill built, and built others very sumptuous. It seemed vnto him that the city of Mexico had too litle water, and that the lake was very muddy, and therefore hee resolved to let in a great course of water, which they of Cuyoacan vsed. For this cause he called the chiefe man of the cittie vnto him, being a famous sorcerer; having propounded his meaning vnto him, the sorcerer wished him to be well advised what hee did, being a matter of great difficulty, and that hee vnderstoode, if he drew the river out of her ordinary course, making it passe to Mexico, hee would drowne the citty. The king supposed these excuses were but to frustrate the effect of his designe, being therefore in choler, he dismissed him home; and a few dayes after hee sent a provost to Cuyoacan, to take this sorcerer: who, having understanding for what intent the king's officers came, he caused them to enter his house, and then he presented himself vnto them in the forme of a terrible eagle, wherewith the provost and his companions being terrified, they returned without taking him. Autzol, incensed herewith, sent others, to whome hee presented himselfe in forme of a furious tygre, so as they durst not touch him. The third came, and they found him in the forme of a horrible serpent, whereat they were much afraide. The king mooved the more with these dooings, sent to tell them of Cuyoacan, that if they brought not the sorcerer bound vnto him, he would raze their citty. For feare whereof, or whether it were of his owne free will, or being forced by the people, he suffered himselfe to be led to the kinge, who presently caused him to be strangled, and then did he put his resolution in practise, forcing a chanell whereby the water might passe to Mexico, whereby hee brought a great current of water into the lake, which they brought with great ceremonies and superstitions, having priests casting incense along the banks, others sacrificed quailes, and with the bloud of them sprinckled the channell bankes, others sounding of cornets, accompanied the water with their musicke. One of the chiefe went attired in a habite like to their goddesse of the water, and all saluted her, saying, that shee was welcome. All which things are painted in the Annales of Mexico: which booke is now at Rome in the holy library, or Vaticane, where a father of our company, that was come from Mexico, did see it, and other histories, the which he did expound to the keeper of his Holinesse library, taking great delight to vnderstand this booke, which before hee could never comprehend. Finally, the water was brought to Mexico, but it came in such aboundaunce, that it had welneere drowned the cittie, as was foretold: and in effect it did ruine a great parte thereof, but it was presently prevented by the industry of Autzol, who caused an issue to be made to draw foorth the water: by meanes whereof hee repaired the buildings that were fallen, with an exquisite worke, being before but poore cottages. Thus he left the citty invironed with water, like another Venice, and very well built: he raigned eleven yeares, and ended with the last and greatest successor of all the Mexicans.

CHAP. xx.—*Of the election of great Monteçuma, the last King of Mexico.*

When the Spaniards entered new Spaine, being in the yeare of our Lorde one thousand five hundred and eighteen,

Monteçuma, second of that name, was the last king of the Mexicaines; I say the last, although they of Mexico, after his death, chose another king, yea, in the life of the same Monteçuma, whom they declared an enemy to his country, as we shall see hereafter. But he that succeeded him, and hee that fell into the hands of the Marquis del Valle,[1] had but the names and titles of Kings, for that the kingdome was in a maner al yeelded to the Spaniards: so as with reason we account Monteçuma for the last king, and so hee came to the periode of the Mexicaine's power and greatnesse, which is admirable, being happened among Barbarians: for this cause, and for that this was the season that God had chosen to reveale vnto them the knowledge of his Gospel, and the kingdome of Iesus Christ, I will relate more at large the actes of Monteçuma, then of the rest.

Before he came to be king, he was by disposition very grave and stayed, and spake little, so as when he gave his opinion in the privy counsell, whereat he assisted, his speeches and discourses made every one to admire him, so as even then he was feared and respected. He retired himselfe usually into a Chappell, appointed for him in the Temple of Vitzilipuztli, where they said their Idoll spake vnto him; and for this cause hee was helde very religious and devout. For these perfections then, being most noble and of great courage, his election was short and easie, as a man upon whom al men's eyes were fixed, as woorthy of such a charge. Having intelligence of this election, hee hidde himselfe in this chappell of the Temple, whether it were by iudgement (apprehending so heavy and hard a burthen as to govern such a people), or rather, as I believe, through hypocrisie, to show that he desired not Empire. In the end they found him, and led him to the place of councell, whither they accompanied him with all possible ioy. Hee marched with such a gravitie, as they all sayd the name of Montecuma agreed very wel with his nature, which is as

[1] Hernan Cortes.

SPEECH OF THE KING OF TEZCUCO.

much to say, an angry Lord. The electors did him great reverence, giving him notice that hee was chosen king: from thence he was ledde before the harth of their gods, to give incense, where he offered sacrifices in drawing bloud from his eares, and the calves of his legges, according to their custome. They attired him with the royall ornaments, and pierced the gristle of his nostrils, hanging thereat a rich emerald, a barbarous and troublous custom, but the desire of rule made all paine light and easy. Being seated in his throne, he gave audience to the Orations and Speeches that were made vnto him, which, according vnto their custome were eloquent and artificiall. The first was pronounced by the King of Tezcuco, which, being preserved, for that it was lately delivered, and very worthy to be heard, I will set it downe word by word, and thus hee sayde: "The concordance and vnitie of voyces upon thy election, is a sufficient testimonie (most noble yong man) of the happines the realme shall receive, as well deserving to be commaunded by thee, as also for the generall applause which all doe show by means thereof. Wherein they have great reason, for the Empire of Mexico doth alreadie so farre extend it selfe, that to governe a world, as it is, and to beare so heavy a burthen, it requires no lesse dexteritie and courage, than that which is resident in thy firm and valiant heart, nor of lesse wisedome and iudgement than thine. I see and know plainely, that the mightie God loveth this Cittie, seeing he hath given vnderstanding to choose what was fit. For who will not believe that a Prince, who before his raigne had pierced the nine vaultes of heaven, should not likewise nowe obtaine those things that are earthlie to releeve his people, aiding himselfe with his best iudgement, being thereunto bound by the dutie and charge of a king. Who will likewise beleeve that the great courage which thou hast alwaies valiantly showed in matters of importance, shuld now faile thee in matters of greatest need? Who will not perswade himselfe but the Mexicaine Empire is come to the height of their soveraignetie, seeing the Lorde of things created hath imparted so great graces vnto thee, that with thy looke onelie thou breedest admiration in them that beholde thee? Rejoice, then, O happy land, to whom the Creator hath given a Prince, as a firme pillar to support thee, which shall be thy father and thy defence, by whom thou shalt be succoured at neede, who wil be more than a brother to his subiects, for his pietie and clemencie. Thou hast a king, who in regard of his estate is not inclined to delights, or will lie stretched out upon his bed, occupied in pleasures and vices; but contrariwise in the middest of his sweete and pleasant sleepe, hee will sodainely awake, for the care he must have over thee, and will not feele the taste of the most savourie meates, having his spirites transported with the imagination of thy good. Tell mee, then (O happy realme), if I have not reason to say that thou oughtest reioyce, having found such a King. And thou noble yong man, and our most mightie Lorde, be confident, and of good courage, that seeing the Lorde of things created hath given thee this charge, hee will also give thee force and courage to mannage it: and thou maiest well hope, that he which in times past hath vsed so great bountie towardes thee, wil not now denie thee his greater gifts, seeing he hath given thee so great a charge, which I wish thee to enioy manie yeares. King Monteçuma was very attentive to this Discourse, which, being ended, they say he was so troubled, that indevouring thrice to answer him, hee could not speake, being overcome with teares, which ioy and content doe vsually cause, in signe of great humilitie. In the end, being come to himselfe, he spake briefly, "I were too blinde, good king of Tezcuco, if I did not know, that what thou hast spoken vnto me, proceeded of meere favour, it pleaseth you to show me, seeing among so manie noble and valiant men within this realme, you have made choise of the least sufficient: and in trueth, I find myself so incapable of a charge of so great importance, that I know not what to doe, but to beseech the Creator of all created things, that he will favour mee, and I intreate you all to pray unto him for me." These words uttered, hee began again to weepe.

CHAP. XXI.—*How Monteçuma ordered the service of his house, and of the warre hee made for his coronation.*

He that in his election made such shew of humilitie and mildenes, seeing himselfe king, beganne presently to discover his aspiring thoughts. The first was, he commaunded that no plebeian should serve in his house, nor beare any royall office, as his predecessours had vsed till then; blaming them that would be served by men of base condition, commaunding that all the noble and most famous men of his realme should live within his pallace, and exercise the offices of his court, and house. Wherevnto an olde man of great authoritie (who had sometimes beene his Schoolemaister) opposed himselfe, advising him to be carefull what hee did, and not to thrust himselfe into the danger of a great inconvenience, in separating himselfe from the vulgare and common people, so as they should not dare to looke him in the face, seeing themselves so reiected by him. He answered, that it was his resolution, and that he would not allow the plebeians to goe thus mingled among the Nobles, as they had doone, saying that the service they did was according to their condition, so as the kings got no reputation, and thus he continued firme in his resolution. Hee presently commanded his counsell to dismisse all the plebeians from their charges and offices, as well those of his houshold as of his court, and to provide knightes, the which was done. After, he went in person to an enterprise necessary for his coronation. At that time a province lying farre off towards the North Ocean was revolted from the crowne, whither he led the flower of his people, well appointed. There he warred with such valour and dexteritie that in the end he subdued all the province, and punished the rebells severely, returning with a great number of captives for the sacrifices and many other spoiles. All the citties made him solemne receptions at his returne, and the Lords thereof gave him water to wash, performing the offices of servants, a thing not vsed by any of his predecessors. Such was the feare and respect they bare him. In Mexico they made the feasts of his coronation with great preparations of dances, comedies, banquets, lights, and other inventions for many daies. And there came so great a wealth of tributes from all his countries that strangers vnknowne came to Mexico, and their very enemies resorted in great numbers disguised to see these feasts, as those of Tlascala and Mechoacan: the which Monteçuma having discovered, he commanded they should be lodged and gently intreated, and honoured as his own person. He also made them goodly galleries like vnto his owne, where they might see and behold the feasts. So they entred by night to those feasts, as the king himselfe, making their sportes and maskes. And for that I have made mention of these provinces, it shall not be from the purpose to vnderstand that the inhabitants of Mechoacan, Tlascala, and Tepeaca, would never yeelde to the Mexicans, but did alwaies fight valiantly against them; yea, sometimes the Mechoacans did vanquish the Mexicans, as also those of Tepeaca did. In which place the Marquis Don Fernando Cortes, after that he and the Spaniards were expelled Mexico, pretended to build their first cittie, the which he called (if I remember rightly) Segura de la Frontera. But this peopling continued little: for having afterwards reconquered Mexico, all the Spaniards went to inhabite there. To conclude, those of Tepeaca, Tlascala, and

Mechoacan have beene alwaies enemies to the Mexicans, although Monteçuma said vnto Cortes that he did purposely forbeare to subdue them, to have occasion to exercise his men of warre, and to take numbers of captives.

CHAP. XXII.—*Of the behaviour and greatnes of Monteçuma.*

This King laboured to be respected, yea, to be worshipped as a god. No Plebeian might looke him in the face; if he did, he was punished with death: he did never set his foote on the ground, but was alwaies carried on the shoulders of Noblemen; and if he lighted, they laid rich tapestry whereon he did go. When he made any voyage, hee and the Noblemen went as it were in a parke compassed in for the nonce, and the rest of the people went without the parke, invironing it in on every side; hee never put on a garment twice, nor did eate or drinke in one vessell or dish above once; all must be new, giving to his attendants that which had once served him: so as commonly they were rich and sumptuous. He was very carefull to have his lawes observed. And when he returned victor from any warre, he fained sometimes to go and take his pleasure, then would he disguise himselfe, to see if his people (supposing if he weare absent) would omitte any thing of the feast or reception. If there were any excesse or defect, he then did punish it rigorously. And also to discerne how his ministers did execute their offices, he often disguised himselfe, offering giftes and presents to the iudges, provoking them to do iniustice. If they offended, they were presently punished with death, without remission or respect, were they Noblemen or his kinsmen; yea, his owne brethren. He was little conversant with his people, and seldome seene, retyring himselfe most commonly to care for the government of his realme. Besides that hee was a great iusticier and very noble, hee was very valiant and happy, by meanes whereof hee obtained great victories, and came to this greatnes, as is written in the Spanish histories, whereon it seemes needelesse to write more. I will onely have a care heereafter to write what the bookes and histories of the Indies make mention of, the which the Spanish writers have not observed, having not sufficiently vnderstood the secrets of this country, the which are things very worthy to be knowne, as we shall see heereafter.

CHAP. XXIII.—*Of the presages and strange prodigies which happened in Mexico before the fall of their Empire.*

Although the holy Scripture forbids vs to give credite to signes and vaine prognostications, and that S. Ierome doth admonish vs not to feare tokens from heaven, as the Gentiles do: yet the same Scripture teacheth vs that monstrous and prodigious signes are not altogether to bee contemned, and that often they are fore-runners of some generall changes and chasticements which God will take, as Eusebius notes well of Cesarea. For that the same Lord of heaven and earth sendes such prodiges and new things in heaven, in the elements, in beasts, and in his other creatures, that this might partly serve as an advertisement to men, and to be the beginning of the paine and chastisement, by the feare and amazement they bring. It is written in the second booke of Macabees that before that great change and persecution of the people of Israel, which was caused by the tyranny of Antiochus, surnamed Epiphanes, whome the holy Scriptures call the root of sinne, there were seene for forty dayes together thorowout all Ierusalem great squadrons of horsemen in the ayre, who with their armour guilt, their lances and targets, and vppon furious horses, with their swordes drawne did strike, skirmish and incounter one against the other: and they say that the inhabitants of Ierusalem seeing this, they prayed to our Lord to appease his wrath, and that these prodegies might turne to good. It is likewise written in the booke of Wisedome, That when God would drawe his people out of Egypt, and punish the Egyptians, some terrible and fearefull visions appeared vnto them, as fires seene out of time in horrible formes. Ioseph in his booke of the Iewish warres sheweth many and great wonders going before the destruction of Ierusalem, and the last captivitie of his wicked people, whome God iustly abhorred: and Eusebius of Cesarea, with others, alleadge the same texts, authorizing prognostications. The Histories are full of like observations in great changes of states and commonweales, as Paulus Orosius witnesseth of many: and without doubt this observation is not vaine nor vnprofitable; for although it be vanitie, yea, superstition, forbidden by the lawe of our God, lightly to beleeve these signes and tokens, yet in matters of great moment, as in the changes of nations, kingdoms, and notable laws, it is no vaine thing, but rather certaine and assured, to beleeve that the wisdome of the most High dooth dispose and suffer these things, foretelling what shoulde happen, to serve (as I have saide) for an advertisement to some and a chasticement to others, and as a witnes to all, that the king of heaven hath a care of man: who as he hath appointed great and fearefull tokens of that great change of the world, which shall bee the day of iudgement, so doth it please him to send wonderful signes to demonstrate lesser changes in divers partes of the world, the which are remarkable, whereof he disposeth according to his eternall wisdome. Wee must also vnderstand that although the divell be the father of lies, yet the King of Glorie makes him often to confesse the trueth against his will, which hee hath often declared for very feare, as hee did in the desart by the mouth of the pos-

LIB. VII.
1 Mac. i.

Sap. vii.

Euseb., lib. i, de eccles. hist.

Mat. i.
Luke iv.

sessed, crying, that Iesus was the Saviour come to destroy him, as he did by the Pythoness, who saide that Paul preached the true God, as when he appeared and troubled Pilate's wife, whom he made to mediate for Iesus a iust man. And as many other histories besides the holy Scripture gave diverse testimonies of idols, in approbation of christian religion, wherof Lactantius, Prosperus, and others make mention. Let them reade Eusebius in his bookes of the preparation of the Gospel, and those of his demonstrations where he doth amply treate of this matter. I have purposely spoken this, that no man should contemne what is written in the Histories and Annales of the Indies touching presages and strange signes, of the approching end and ruine of their kingdome, and of the Divelles tyranny, whom they worshipped altogether. Which in my opinion is worthy of credite and beliefe, both for that it chanced late, and the memory is yet fresh, as also for that it is likely that the Divell lamented at so great a change, and that God by the same meanes beganne to chastice their cruell and abominable idolatries. I will therefore set them downe heere as true things. It chanced that Monteçuma having raigned many yeers in great prosperity, and so pufft vp in his conceit, as hee caused himselfe to be served and feared, yea, to be worshipped as a god, that the Almighty Lord beganne to chastice him, and also to admonish him, suffering even the very Divelles whome he worshipped to tell him these heavy tidings of the ruine of his kingdome, and to torment him by visions, which had never bin seen; wherewith hee remained so melancholy and troubled, as he was voyde of iudgement. The idoll of those of Cholula, which they called Quetzalcoatl, declared that a strange people came to possesse his kingdomes. The king of Tezcuco (who was a great Magitian, and had conference with the Divell) came one day at an extraordinarie houre to visite Monteçuma, assuring him that his gods had tolde him that there were great losses preparing for him and for his whole realme: many witches and sorcerers went and declared as much; amongst which there was one did very particularly foretell him what should happen: and as he was with him hee tolde him that the pulses of his feete and hands failed him. Monteçuma, troubled with these news, commanded all those sorcerers to be apprehended: but they vanished presently in the prison, wherewith hee grewe into such a rage, that hee might not kill them, as hee putte their wives and children to death, destroying their houses and families. Seeing himselfe importuned and troubled with those advertisements, he sought to appease the anger of his gods: and for that cause hee laboured to bring a huge stone, thereon to make great sacrifices. For the effecting whereof hee sent a great number of people with engines and instruments to bring it: which they could by no meanes moove, although (being obstinate) they had broken many instruments. But as they strove still to raise it they heard a voyce ioyning to the stone, which said they laboured in vaine, and that they should not raise it, for that the Lorde of things created would no more suffer those things to be doone there. Monteçuma, vnderstanding this, commaunded the sacrifice to be perfourmed in that place, and they say the voyce spake againe: "Have I not told you that it is not the pleasure of the Lord of things created that it should be done: and that you may well know that it is so, I will suffer my selfe to be transported a little, then after you shall not moove mee". Which happened so indeede; for presently they carried it a small distance with great facility, then afterwards they could not moove it, till that after many prayers it suffered it selfe to be transported to the entry of the citty of Mexico, where sodainly it fel into the Lake, where, seeking for it, they could not finde it, but it was afterwards found in the same place from whence they had remooved it, wherewith they remayned amazed and confounded. At the same time there appeared in the heavens a great flame of fire, very bright, in the forme of a Pyramide, which beganne to appeare at midnight, and went still mounting vntill the Sunne rising in the morning, where it stayed at the South, and then vanished away. It shewed it self in this sort the space of a whole yeare, and ever as it appeared the people cast foorth great cries as they were accustomed, beleeving it was a presage of great misfortune. It happened also that fire tooke the Temple, whenas no body was within it, nor neare vnto it, neither did there fall any lightning or thunder: wherevpon the guardes crying out, a number of people ran with water, but nothing could helpe, so as it was all consumed; and they say the fire seemed to come forth of peeces of timber, which kindled more by the water that was cast vpon it. There was a Comet seene in the day time, running from the west to the east, casting an infinite number of sparkles, and they say the forme was like to a long taile, having three heads.

The great lake betwixt Mexico and Tezcuco, without any wind, earthquake, or any other apparent signe, beganne sodainely to swell, and the waves grewe in such sort, as all the buildings neare vnto it fell downe to the ground. They say at that time they heard many voices, as of a woman in paine, which sayde sometimes, "O my children, the time of your destruction is come", and otherwhiles it sayde, "O my children, whither shall I carry you, that you perish not utterly?" There appeared, likewise, many monsters with two heads, which, being carried before the king, sodainely vanished. There were two that exceeded all other monsters, being very strange; the one was, the fishers of the lake tooke a bird as bigge as a crane, and of the same colour, but of a strange and vnseene form. They caried it to Monteçuma, who at that time was in the pallace of tears and mourning, which was all hanged with blacke, for as he had many palaces for his recreation, so had he also others for times of affliction, wherewith hee was then heavily charged and tormented, by reason of the threatnings his gods had given him by these sorrowfull advertisements. The fishers came about noone, setting this bird before him, which had on the toppe of his head a thing bright and transparent, in forme of a looking glasse, wherein he did behold a warre-like nation comming from the east, armed, fighting, and killing. He called his Divines and Astronomers (whereof there was a great number), who, having seen these things, and not able to yeelde any reason of what was demaunded of them, the bird vanished away, so as it was never more seene: wherevpon Monteçuma remained very heavy and sorrowfull. The other which happened was a labourer, who had the report of a very honest man, came vnto him, telling him, that being the day before at his worke, a great Eagle flew towardes him, and tooke him vppe in his talents, without hurting him, carying him into a certaine cave, where it left him; the Eagle pronouncing these words, "Most mightie Lorde, I have brought him whome thou hast commaunded me". This Indian labourer looked about on every side, to whome hee spake, but hee sawe no man. Then he heard a voyce which sayde vnto him, " Doost thou not knowe this man, whome thou seest lying vpon the ground"; and looking thereon, he perceived a man lie very heavy asleepe, with royall ensignes, floures in his hand, and a staffe of perfumes burning, as they are accustomed to vse in that countrey, whome the labourer beholding, knew it was the great king Monteçuma, and answered presently: "Great Lorde, this resembles our King Monteçuma." The voyce saide againe, "Thou saiest true, behold what he is, and how he lies asleepe, carelesse of the great miseries and afflictions prepared for him. It is nowe time that he pay the great number of offences hee hath doone to God, and that he receive the punishment of his tyrannies and great pride, and yet thou seest how carelesse hee lies, blinde in his owne miseries, and without any feeling. But to the end thou maiest the better see him, take the staffe of perfumes hee holdes burning in his hand, and put it to his face, thou shalt then find him without feeling." The poore laborer durst not approach neere him, nor doe as he was commaunded, for the great feare they all hadde of this king. But the voyce saide, "Have no feare, for I am without comparison greater than this King, I can destroy him, and defend him, doe therefore what I commaund thee." Wherevpon the laborer took the staffe of perfumes out of the king's hand, and put it burning to his nose, but he mooved not, nor showed any feeling.

This done, the voice said vnto him, that seeing he had found the king so sleepy, he should go awake him, and tell him what he had seene. Then the Eagle, by the same commandment, tooke the man in his talents, and set him in the same place where he found him, and for accomplishment of that which it had spoken, hee came to advertise him. They say, that Monteçuma looking on his face, found that he was burnt, the which he had not felt till then, wherewith he continued exceedingly heavy and troubled. It may be, that what the laborer reported, had happened vnto him by imaginary vision. And it is not incredible, that God appointed by the meanes of a good Angell, or suffered by a bad, that this advertisement should be given to the labourer for the king's chasticement, although an infidell, seeing that we read in the Holy Scriptures, that infidells and sinners have had the like apparitions and revelations, as Nabucadonosor, Balaam, and the Pithoness of Saul. And if some of these apparitions did not so expresly happen, yet, without doubt, Monteçuma had many great afflictions and discontentments, by reason of sundry and divers revelations which he had, that his kingdome and law should soon end.

Chap. XXIV.—*Of the newes Monteçuma received of the Spaniards arrival in his Country, and of the Ambassage he sent them.*

In the fourteenth yeare of the raigne of Monteçuma, which was in the yeare of our Lord 1517, there appeared in the North seas, shippes, and men landing, whereat the subiects of Monteçuma wondred much, and desirous to learne, and to be better satisfied what they were, they went aboord in their canoes, carrying many refreshings of meats and stuffes to make apparrell, vpon colour to sell them. The Spaniards received them into their shippes, and in exchange of their victualls and stuffes, which were acceptable vnto them, they gave them chaines of false stones, red, blew, greene, and yellow, which the Indians imagined to be precious stones. The Spaniards informing themselves who was their king, and of his great power, dismissed them, willing them to carry those stones vnto their lord, saying, that for that time they could not goe to him, but they would presently returne and visit him. Those of the coast went presently to Mexico with this message, carrying the representation of what they had seene painted on a cloth, both of the shippes, men, and stones which they had given them. King Monteçuma remained very pensive with this message, commanding them not to reveale it to any one. The day following, he assembled his counsell, and having showed them the painted clothes and the chaines, he consulted what was to be done; where it was resolved to set good watches vpon all the sea coastes, to give present advertisement to the king of what they should discover. The yeare following, which was in the beginning of the yeare 1518, they discovered a fleet at sea, in the which was the Marques del Valle Don Fernando Cortes, with his companions, a newes which much troubled Monteçuma, and conferring with his counsell, they all said, that without doubt, their great and antient Lord Quetzalcoatl was come, who had saide, that he would returne from the East, whither he was gone. The Indians held opinion, that a great Prince had in times past left them, and promised to returne. Of the beginning and ground of which opinion shall be spoken in another place. They therefore sent five principall Ambassadors with rich presents, to congratulate his comming, saying, they knewe well that their great Lord Quetzalcoatl was come, and that his servant Monteçuma sent to visit him, for so hee accounted himselfe. The Spaniards vnderstood this message by the meanes of Marina, an Indian woman whom they brought with them, and vnderstood the Mexicane tongue. Fernando Cortes finding this a good occasion for his entry, commanded to deck his chamber richly, and being set in great state and pompe, he caused the Ambassadors to enter, who omitted no showes of humilitie, but to worshippe him as their god.

They delivered their charge, saying, that his servant Monteçuma sent to visit him, and that he held the country in his name as his lievetenant; that he knew well it was the Topilcin which had beene promised them many yeares since, who should returne again vnto them. And therefore they brought him such garments as he was wont to weare, when hee did converce amongst them, beseeching him to accept willingly of them, offering him many presents of great value. Cortes receiving the presents, answered that he was the same they spake of, wherewith they were greatly satisfied, seeing themselves to be curteously received and intreated by him (for in that, as wel as in other things, this valiant captaine deserved commendations); that if this course had been continued, to win them by love, it seemed the best occasion was offered that might be devised, to draw this country to the Gospel by peace and love: but the sinnes of these cruel homicides and slaves of Satan required punishment from heaven, as also those of many Spaniards, which were not in small number. Thus the high iudgements of God disposed of the health of this nation, having first cut off the perished rootes: and as the Apostle saieth, the wickednes and blindenes of some, hath beene the salvation of others. To conclude, the day after this Ambassage, all the Captaines and Commanders of the fleete came vnto the Admirall, where vnderstanding the matter, and that this realme of Monteçuma was mightie and rich: it seemed fit to gaine the reputation of brave and valiant men among this people, and that by this meanes (although they were few), they should bee feared and received into Mexico. To this end they discharged all their artillerie from their shippes, which being a thing the Indians had never heard, they were amazed, as if heaven had fallen vpon them. Then the Spaniards beganne to defie them to fight with them: but the Indians not daring to hazard themselves, they did beate them and intreate them ill, showing their swordes, lances, partisans, and other armes, wherewith they did terrifie them much. The poore Indians were by reason heereof so fearefull and amazed, as they changed their opinion, saying, that their Lord Topilcin came not in this troup. But they were some gods (their enemies), come to destroy them. Whenas the Ambassadors returned to Mexico, Monteçuma was in the house of audience; but before he would heare them, this miserable man commanded a great number of men to be sacrificed in his presence, and with their blood to sprinkle the Ambassadors, supposing by this ceremony (which they were accustomed to do in solemne Ambassages), to receive a good answer. But vnderstanding the report and information of the maner of their shippes, men, and armes, he stoode perplexed and confounded: then taking counsell thereon, he found no better meanes then to labour to stoppe the entrie of these strangers by coniurations and magicke Artes. They had accustomed often to vse this meanes, having great conference with the divell, by whose helpe they sometimes obtained strange effects. They therefore assembled together all the Sorcerers, Magicians, and Inchanters, who being perswaded by Monteçuma, they tooke it in charge to force this people to returne vnto their country. For this consideration, they went to a certaine place which they thought fit for the invocation of their divells, and practising their artes (a thing worthy of consideration), they wrought all they could; but seeing nothing could prevaile against the Christians, they went to the king, telling him that they were more than men, for that nothing might hurt them, notwithstanding all their coniurations and inchantments. Then Monteçuma advised him of another pollicie, that faining to be very well contented with their comming, he commanded all his countries to serve these celestiall gods that were come into his land. The whole people was in great heavinesse and amazement, and often newes came that the Spaniards inquired for the King, of his manner of life, of his house and meanes. He was exceedingly vexed herewith; some of the people and other Necromancers advised him to hide himselfe, offering to place him whereas no creature should ever finde him. This seemed base vnto him, and therefore he resolved to attend them, although it were dying. In the end he left his houses and royall pallaces to lodge in others, leaving them for these gods as he said.

Chap. xxv.—*Of the Spaniards entrie into Mexico.*

I pretende not to intreate of the acts and deedes of the Spaniards who conquered New Spaine, nor the strange adventures which happened vnto them, nor of the courage and invincible valour of their Captaine Don Fernando

MONTEZUMA'S STRATEGY.

Cortes: for that there are many histories and relations thereof, as those which Fernando Cortes himselfe did write to the Emperour Charles the fift, although they be in a plaine stile and farre from arrogancie, the which doe give a sufficient testimony of what did passe, wherein he was worthy of eternall memory, but onely to accomplish my intention. I am to relate what the Indians report of this action, the which hath not to this day lyɔene written in our vulgar tong. Monteçuma therefore, having notice of this Captaines victories, that he advanced for his conquest, that hee was confederate and ioyned with them of Tlascala, his capitall enemies, and that he had severely punished them of Cholula his friends, he studied how to deceive him, or else to try him in sending a principall man vnto him, attyred with the like ornaments and royall ensignes, the which shuld take vpon him to be Monteçuma, which fiction being discouered to the Marquis by them of Tlascala (who did accompany him), he sent him backe, after a milde and gentle reprehension, in seeking so to deceive him: wherevpon Monteçuma was so confounded, that for the feare thereof, he returned to his first imaginations and practises, to force the christians to retyre, by the invocation of coniurers and witches. And therefore he assembled a greater number then before, threatning them that if they returned without effecting what he had given them in charge, not any one should escape, wherevnto they all promised to obey. And for this cause all the divells officers went to the way of Chalco, by the which the Spaniards should passe, when, mounting to the top of a hill, Tezcatlipuca, one of their principall gods, appeared vnto them, as comming from the Spaniards camp, in the habite of Chalcas, who had his breast bound about eight folde with a corde of reeds, hee came like a man beside himselfe, out of his wits, and drunke with rage and furie. Being come to this troupe of witches and coniurers, he staied, and spake to them in great choller, "Why come you hither: what doth Monteçuma pretend to doe by your meanes? He hath advised himselfe too late: for it is now determined that his Kingdom and honour shall be taken from him, with all that he possesseth, for punishment of the great tyrannies he hath committed against his subjects, having governed not like a Lord, but like a traitour and tyrant." The inchanters and coniurers, hearing these words, knew it was their idoll, and, humbling themselves before him, they presently built him an altar of stone in the same place, covering it with flowers which they gathered thereaboutes, but he contrariwise, making no account of these things, beganne againe to chide them, saying, "What come you hither to do, O yee traitours? Returne presently and behold Mexico, that you may vnderstand what shall become thereof". And they say that, turning towards Mexico to behold it, they did see it flaming on fire. Then the divell vanished away, and they, not daring to passe any farther, gave notice thereof to Monteçuma, whereat he remained long without speaking, looking heavily on the ground; then he said, What shall we doe if god and our friends leave vs, and contrariwise, they helpe and favour our enemies? I am now resolute, and we ought all to resolve in this point, that happen what may, we must not flie nor hide ourselves, or shew any signe of cowardice. I onely pittie the aged and infants, who have neither feete nor hands to defend themselves. Having spoken this, he held his peace, being transported into an extasie. In the end the Marquis approaching to Mexico, Monteçuma resolved to make of necessitie a vertue, going three or foure leagues out of the cittie to receive him with a great maiesty, carried vpon the shoulders of foure Noblemen, vnder a rich canopie of gold and feathers: when they mette, Monteçuma discended, and they saluted one another very curteously. Don Fernando Cortes said vnto him that he should not care for any thing, and that he came not to take away his realme, nor to diminish

INTERVIEW WITH MONTEZUMA.

his authoritie. Monteçuma lodged Cortes and his companions in his royall pallace, the which was very stately, and he himselfe lodged in other private houses. This night the souldiers for ioy discharged their artillery, wherewith the Indians were much troubled, being vnaccustomed to heare such musicke. The day following Cortes caused Monteçuma and all the Nobles of his Court to assemble in a great hall, where, being set in a high chaire, he said vnto them that hee was servant to a great prince, who had sent them into these countries to doe good workes, and that having found them of Tlascala to be his friendes (who complained of wrongs and greevances done vnto them daily by them of Mexico), he would vnderstand which of them was in the blame, and reconcile them, that heereafter they might no more afflict and warre one against another: and in the meane time he and his bretheren (which were the Spaniards) would remaine still there without hurting them : but contrariwise, they would helpe them all they could. He laboured to make them all vnderstand this discourse, vsing his interpreters and truchmen. The which being vnderstoode by the King and the other Mexicane Lords, they were wonderfully well satisfied, and shewed great signes of love to Cortes and his company. Many hold opinion that if they had continued the course they began that day, they might easily have disposed of the king and his kingdome, and given them the law of Christ without any great effusion of bloud. But the iudgements of God are great, and the sins of both parties were infinite: so as not having followed this course, the busines was deferred : yet in the end God shewed mercy to this nation, imparting vnto them the light of his holy Gospel, after he had shewed his iudgement, and punished them that had deserved it, and odiously offended his divine reverence. So it is that by some occasions many complaints, griefs, and icalosies grew on either side. The which Cortes finding, and that the Indians mindes began to be distracted from them, he thought it necessary to assure himself, in laying hand vpon king Monteçuma, who was seazed on, and his legs fettered. Truly this act was strange vnto all men, and like vnto that other of his, to have burnt his ships, and shut himselfe in the midst of his enemies, there to vanquish or to die. The mischiefe was, that by reason of the vnexpected arrival of Pamphilo de Narvaez at Vera Cruz, drawing the country into mutiny, Cortes was forced to absent himselfe from Mexico, and to leave poor Monteçuma in the handes of his companions, who wanted discretion, nor had not moderation like vnto him; so as they grew to that discention, as there was no meanes to pacifie it.

Chap. xxvi.—*Of the death of Monteçuma, and the Spaniards departure out of Mexico.*

Whenas Cortes was absent from Mexico, he that remained his lievetenant resolved to punish the Mexicans severely, causing a great number of the nobilitie to be slaine at a maske which they made in the pallace, the which did so far exceede, as all the people mutinied, and in a furious rage took armes to be revenged and to kil the Spaniards. They therefore besieged them in the pallace, pressing them so neere, that all the hurt the Spaniards could do them with their artillery and crosse-bowes, might not terrifie them, nor force them to retyre from their enterprise, where they continued many daies, stopping their victualls, nor suffering any one to enter or issue forth. They did fight with stones, and cast dartes after their maner, with a kind of lances like vnto arrowes, in the which there are foure or six very sharpe rasors, the which are such (as the histories report) that in these warres an Indian with one blow of these rasors almost cut off the necke of a horse; and as they did one day fight with this resolution and furie, the Spaniards, to make them cease, shewed forth Monteçuma, with another of the chiefe Lords of Mexico, vpon the top of a platform of the house, covered with the targets of two souldiers that were with them. The Mexicanes, seeing their Lord Monteçuma, staied with great silence. Then Monteçuma caused the Lord to advise them to pacifie themselves, and not to warre against the Spaniards, seeing that (hee being a prisoner) it could little profite him. The which being vnderstood by a yong man called Quicuxtemoc, whom they now resolved to make their king, spake with a loud voice to Monteçuma, willing him to retyre like a villaine, that seeing he had bin such a coward as to suffer himselfe to be taken, they were no more bound to obey him, but rather should punish him as he deserued, calling him woman for the more reproach, and then hee beganne to draw his bowe and to shoote at him, and the people beganne to cast stones at him, and to continue their combate. Many say that Monteçuma was then hurt with a stone, whereof he died. The Indians of Mexico affirme the contrarie, and that he died as I will shew hereafter. Alvarado and the rest of the Spaniards, seeing themselves thus pressed, gave intelligence to Captaine Cortes of the great danger they were in : who having with an admirable dexteritie and valour given order to Narvaez affaires, and assembled the greatest part of his men, he returned with all speede to succour them of Mexico, where observing the time the Indians rest (for it was their custom in war to rest every fourth day:) He one day advanced with great policy and courage, so as both he and his men entred the pallace, whereas the Spaniards had fortified themselves: they then shewed great signes of ioy in discharging their artillery. But as the Mexicans furie increased (being out of hope to defend themselves) Cortes resolved to passe away secretly in the night without bruite. Having therefore made bridges to passe two great and dangerous passages, about midnight they issued forth as secretly as they could, the greatest part of his people having passed the first bridge, they were discovered by an Indian woman before they could passe the second, who cried out their enemies fled, at the which voice all the people ran together with a horrible furie: so as in passing the second bridge, they were so charged and pursued, as there remained above three hundred men slaine and hurt in one place; where at this day there is a smal hermitage, which they vnproperly cal of Martyrs. Many Spaniards (to preserve the gold and iewells which they had gotten), perished, and others staying to carry it away, were taken by the Mexicans, and cruelly sacrificed to their idols. The Mexicans found king Monteçuma dead, and wounded as they say with poiniards, and they hold opinion that that night the Spaniards slew him with other Noblemen. The Marquis in his relation sent to the Emperour, writes the contrary, and that the Mexicans killed him that night with a son of Monteçuma, which he led with him amongst other noblemen, saying, that all the treasure of gold, stones, and silver fell into the lake and was never more seene. But howsoever, Monteçuma died miserably, and paied his deserts to the iust iudgement of our Lord of heaven for his pride and tyranny: his body falling into the Indians power, they would make him no obsequies of a king, no, not of an ordinarie person, but cast it away in great disdaine and rage. A servant of his having pittie of this king's miserie (who before had bene feared and worshipped as a God) made a fier thereof, and put the ashes in a contemptible place. Returning to the Spaniards that escaped, they were greatly tyred and turmoiled, the Indians following them two or three daies very resolutely, giving them no time of rest, being so distressed for victualls, as a few graines of Mays were divided amongst them for their meate. The relations both of the Spaniards and Indians agree, that God delivered them here miraculously, the Virgin Mary defending them on a little hill, whereat this day, three leagues from Mexico, there is a Church built in remembrance thereof, called our Lady of succour. They retyred to their antient friends of Tlascala, whence, by their aide and the valour and pollicie of Cortes, they returned afterwards to make war against Mexico, by water and land, with an invention of brigantines, which they put into the lake, where, after many combates, and above threescore dangerous battailes, they conquered Mexico, on S. Hippolitus day, the 13 of August 1521. The last king of the Mexicans (having obstinately maintained the wars) was in the end taken in a great canoe, whereinto he fled, who, being brought, with some other of the chiefest noblemen, before Fernando Cortes, this pettie king, with a strange resolution and courage, drawing his dagger, came neere to Cortes, and said vnto him, "Vntill this day I have done my best indevour for the defence of my people: now am I no farther bound, but to give thee this dagger to kill me therewith." Cortes answered, that he would not kill him, neither was it his intention to hurt them: but their obstinate folly was guiltie of all the misery and afflictions they had suffered, neither were they ignorant how often he had required peace and amity at their hands. He then commanded them to be intreated curteously. Many strange and admirable things chanced in this conquest of Mexico: for I neither hold it for an vntruth, nor an addition, which many write, that God favoured the Spaniards by many miracles: for else it had bin impossible to surmount so many difficulties without the favour of heaven, and to subiect this nation with so few men. For although we were sinners, and vnworthy so great a favour, yet the cause of our God, the glorie of our faith, the good of so many thousand soules, as were in these countries, whome the Lord had predestinate, wrought this change which wee now see by supernaturall meanes, and proper to himselfe which calls the blinde and prisoners to the knowledge of himselfe, giving them light and libertie by his holy Gospel. And to the end you may the better vnderstand this, and give credite therevnto, I will aledge some examples which, in my opinion, are fit for this history.

CHAP. XXVII.—*Of some miracles which God hath showed at the Indies, in favour of the faith, beyond the desert of those that wrought them.*

Santa Cruz de la Sierra is a very great province, in the Kingdome of Peru, neighbour to diverse infidell nations, which have not yet any knowledge of the Gospel, if since my departure the fathers of our company which remaine there have not instructed them. Yet this province of Santa Cruz is peopled by Christians, and there are many Spaniards, and great numbers of Indians baptized. The maner how Christianitie entred was thus. A souldier of a lewd life, resident in the province of Charcas, fearing punishment, being pursued for his offences, went farre vp into the countrie, and was received curteously by this barbarous people. The Spaniard seeing them in a great extremity for water, and that to procure raine they vsed many superstitious ceremonies, according to their vsuall maner, he said vnto them, that if they would do as he said, they should presently have raine, the which they willingly offered to performe. Then the souldier made a great crosse, the which he planted on a high and eminent place, commanding them to worship it and to demand water, the which they did. A wonderful thing to see, there presently fel such abundance of raine, as the Indians tooke so great devotion to the holy crosse, as they fled vnto it in all their necessities, and obtained all they demanded: so as they brake downe their idolls, and beganne to carry the crosse for their badge, demanding preachers to instruct and to baptise them. For this reason, the province to this day hath beene called Santa Cruz de la Sierra. But to the end we may see by whom God wrought these miracles, it shall not be vnfit to show how that this souldier, after he had some yeares done these miracles, like an Apostle, and yet nothing reformed in his lewd course of life, left the province of Charcas, and continuing in his wicked courses, was publikely hanged at Potosi. Polo (who knew him wel) writes all this, as a notable thing happened in his time. Cabeça de Vaca, who since was governour of Paraguay, writes what happened vnto him in his strange perigrination in Florida, with two or three other companions, the onely remainder of an army, where they continued ten yeares with these Barbarians, traveling and searching even vnto the South sea, being an author worthy of credite: he saieth, that these Barbarians did force them to cure certaine diseases, threatning them with death if they did it not; they being ignorant in any part of phisicke, and having nothing to apply, forced by necessitie, made evangelicall medicines, saying the praiers of the Church, and making the signe of the crosse, by meanes whereof they cured these diseases, which made them so famous, as they were forced to exercise this office in all townes as they passed, the which were innumerable, wherein our Lord did aide them miraculously, and they themselves were thereat amazed, being but of an ordinarie life; yea, one of them was a Negro. Lancero was a souldier of Peru, of whom they knew no other merit but to be a souldier: he spake certaine good wordes vpon wounds, and making the signe of the crosse, did presently cure them: so as they did say (as in a proverbe), the psalme of Lancero. Being examined by such as held authority in the Church, his office and works were approved. Some men worthy of credite report (and I have heard it spoken), that in the cittie of Cusco, whenas the Spaniards were besieged and so straightly pressed, that without helpe from heaven it was impossible to escape, the Indians casting fire on the tops of the houses, whither the Spaniards were retyred (in which place the great Church is now built),[1] and although the covering were of a kind of straw, which they call Chicho,[2] and that the fire they cast was of very resinous faggots; yet nothing was set on fire, nor burnt, for that there was a woman did quench it presently, the which the Indians did visibly see, as they confessed afterwards being much amazed. It is most certaine, by the relations of many, and by the histories which are written, that in divers battailes which the Spaniards had, as well in New Spaine as in Peru, the Indians their enemies did see a horseman in the aire, mounted on a whit horse, with a sword in his hand, fighting for the Spaniards, whence comes the great reverence they beare at the Indies to the glorious Apostle Saint Iames. Other whiles they did see in some battailes the image of our Ladie, from whom the Christians have received in those partes incomparable favours and benefites: if I should particularly relate all the workes of heaven as they happened, it would make a very long discourse. It sufficeth to have said this, by reason of the favour which the Queene of glorie did to our men when they were pressed and pursued by the Mexicans, the which I have set downe, to the end we may know how our Lord hath had a care to favour the faith and Christian religion, defending those that maintained it, although happily by their workes they deserved not so great favours and benefites from heaven. And therefore we ought not to condemne all these things of the first Conquerours of the Indies, as some religious and learned men have done, doubtlesse with a good zeale, but too much affected. For although, for the most part, they were covetous men, cruell,

[1] The great hall of the palace of Ynca Huiracocha, now the cathedral of Cuzco.   [2] Ychu. (*Stipa Ychu*).

and very ignorant in the course that was to be observed with the Infidels, who had never offended the Christians, yet can we not deny but on their part there was much malice against God and our men, which forced them to vse rigor and chastisement. And, moreover, the Lord of all (although the faithfull were sinners), would favour their cause and partie, even for the good of the Infidells, who should bee converted vnto the holy Gospel by this meanes, for the waies of God are high, and his paths wonderful.

CHAP. XXVIII.—*Of the maner how the Divine Providence disposed of the Indies, to give an entrie to Christian Religion.*

I will make an end of this historie of the Indies, showing the admirable meanes whereby God made a passage for the Gospel in those partes, the which we ought well to consider of, and acknowledge the providence and bountie of the Creator. Every one may vnderstand by the relation and discourse I have written in these bookes, as well at Peru as in New Spaine, whenas the Christians first set footing, that these Kingdomes and Monarchies were come to the height and period of their power. The Yncas of Peru, possessing from the Realme of Chile beyond Quito, which are a thousand leagues, being most aboundant in gold, siluer, and all kinds of riches: as also in Mexico, Monteçuma commaunded from the North Ocean sea vnto the South, being feared and worshipped, not as a man, but rather as a god. Then was it, that the most high Lord had determined that that stone of Daniel, which dissolved the Realmes and Kingdoms of the world, should also dissolve those of this new world. And as the lawe of Christ came whenas the Romane Monarchie was at her greatnes: so did it happen at the West Indies, wherein we see the iust providence of our Lord. For being then in the world, I meane in Europe, but one head and temporall Lord, as the holy Doctors do note, whereby the Gospel might more easily be imparted to so many people and nations. Even so hath it happened at the Indies, where having given the knowledge of Christ to the Monarchs of so many Kingdomes, it was a meanes that afterwards the knowledge of the gospell was imparted to all the people: yea, there is herein a speciall thinge to be observed, that as the Lordes of Cuzco and Mexico conquered new landes, so they brought in their owne language, for although there were (as at this day) great diuersitie of tongues, yet the courtlie speech of Cuzco did, and doth at this day, runne above a thousand leagues, and that of Mexico did not extend farre lesse, which hath not beene of small importance, but hath much profited in making the preaching easie at such a time, when as the preachers had not the gift of many tongues, as in old tymes. He that woulde knowe what a helpe it hath beene for the conversion of this people in these two greate Empyres, and the greate difficultie they haue founde to reduce those Indians to Christ, which acknowledge no Soueraigne Lord, let him goe to Florida, Brasil, the Andes,[1] and many other places, where they have not prevailed so much by their preaching in fiftie yeares, as they have done in Peru and Newe Spaine in lesse than five. If they will impute the cause to the riches of the countrie, I will not altogether denie it. Yet were it impossible to have so great wealth, and to bee able to preserve it, if there had not beene a Monarchie. This is also a worke of God in this age, that we, Preachers of the gospell being so colde and without zeale, Merchants and Soldiers, with the heat of covetousness and desire of command, search and discouer newe people whither wee passe with our commodities. For as Saint Austin saith, the Prophesie of Esaias is fulfilled, in that the Church of Christ is extended,

August., lib. ii, de Conc evangel., xxxvi.

[1] Antis. Not the mountains, but the Peruvian province of Antisuyu, the wild forests to the eastward of the Andes.

not onely to the right hand, but also to the left: which is (as he declareth) by humaine and earthly meanes, which they seeke more commonly than Iesus Christ. It was also a great providence of our Lord, that whenas the first Spaniardes arrived there, they founde ayde from the Indians themselves, by reason of their partialities and greate diuisions.

This is well knowne in Peru, that the division betwixt the two brothers Atahualpa and Huascar, the great King Huayna Ccapac their father being newly dead, gave entry to the Marquis Don Francisco Pizarro, and to the Spaniards, for that either of them desired his alliance, being busied in warre one against the other. The like experience hath beene in New Spaine, that the aide of those of the province of Tlascala, by reason of their continuall hatred against the Mexicaines, gave the victory and siegniory of Mexico to the Marquis Fernando Cortes and his men, and without them it had beene impossible to have wonne it, yea, to have maintained themselves within the country.

They are much deceived that so little esteeme the Indians, and iudge that (by the advantage the Spaniards have over them in their persons, horses, and armes, both offensive and deffensive), they might easily conquer any land or nation of the Indies.

Chile standes yet, or, to say better, Arauco and Tucapel, which are two cities, where our Spaniards could not yet winne one foote of ground, although they have made warre there above five-and-twenty yeares, without sparing of any cost. For this barbarous nation, having once lost the apprehention of horse and shotte, and knowing that the Spaniards fall as well as other men, with the blow of a stone or of a dart, they hazard themselves desperately, entring the pikes vppon any enterprise. How many yeares have they levied men in New Spaine, to send against the Chichimecos, which are a small number of naked Indians, armed onely with bowes and arrowes: yet, to this day, they could not bee vanquished, but contrariwise, from day to day they grow more desperate and resolute. But what shall wee say of the Chunchos, of the Chirihuanos, of the Pilcoçones, and all the other people of the Andes? Hath not all the flower of Peru beene there, bringing with them so great provision of armes and men, as we have seene? What did they? With what victories returned they? Surely they returned very happy in saving of their lives, having lost their baggage and almost all their horses. Let no man thinke (speaking of the Indians), that they are men of nothing; but if they thinke so, let them go and make triall. Wee must then attribute the glory to whom it appertaines, that is, principally to God, and to his admirable providence: for if Monteçuma in Mexico, and the Ynca in Peru, had bin resolute to resist the Spaniards, and to stoppe their entrie, Cortes and Pizarro had prevailed little in their landing, although they were excellent Captaines. It hath also beene a great helpe to induce the Indians to receive the law of Christ, the subiection they were in to their Kings and Lords, and also the servitude and slaverie they were helde in by the divell's tyrannies and insupportable yoke. This was an excellent disposition of the Divine Wisedome, the which drawes profite from ill to a good end, and receives his good from another's ill, which it hath not sowen. It is most certaine that no people of the West Indies have been more apt to receive the Gospel then those which were most subiect to their Lords, and which have beene charged with the heaviest burthens, as well of tributes and services, as of customes and bloodie practises. All that which the Mexicane Kings, and those of Peru did possesse, is at this day most planted with Christian religion, and where there is least difficultie in the government and ecclesiasticall discipline. The Indians were so wearied with the heavy and insupportable yoke of

Sathan's lawes, his sacrifices and ceremonies, whereof wee have formerly spoken, that they consulted among themselves to seeke out a new law, and another God to serve. And therefore the law of Christ seemed vnto them, and doth at this day seeme iust, sweete, clean, good, and full of happinesse.

And that which is difficult in our law, to beleeve so high and soveraigne Misteries, hath beene easy among them, for that the Divell had made them comprehend things of greater difficultie, and the self-same things which he had stolen from our Evangelicall law, as their maner of communion and confession, their adoration of three in one, and such other like, the which, against the will of the enemy, have holpen for the easie receiving of the truth by those who before had imbraced lies. God is wise and admirable in all his works, vanquishing the adversarie even with his owne weapon, hee takes him in his owne snare, and kills him with his owne sword. Finally, our God (who had created this people, and who seemed to have thus long forgot them), when the houre was come, hee would have the same divells, enemies to mankinde, whom they falsely held for gods, should give a testimony against their will, of the true law, the power of Christ, and the triumph of the crosse, as it plainely appeares by the presages, prophesies, signes, and prodiges, heere before mentioned, with many others happened in divers partes, and that the same Ministers of Sathan, Sorcerers, Magitians, and other Indians have confessed it. And we cannot deny it (being most evident and knowne to all the world), that the Divell dareth not hisse, and that the practises, oracles, answers, and visible apparitions, which were so ordinary throughout all this infidelitie, have ceased, whereas the Cross of Christ hath beene planted, where there are Churches, and where the name of Christ hath beene confessed. And if there be at this day any cursed minister of his, that doth participate thereof, it is in caves, and on the toppes of mountaines, and in secret places, farre from the name and communion of Christians. The Soveraigne Lord be blessed for his great mercies, and for the glory of his holy name. And in truth, if they did governe this people, temporally and spiritually, in such sort as the law of Iesus Christ hath set it downe, with a mild yoake and light burthen, and that they would impose no more vppon them then they can well beare, as the letters patents of the good Emperour of happy memorie doe command, and that they would employ halfe the care they have to make profite of these poore men's sweats and labours, for the health of their soules, it were the most peaceable and happy Christian part of all the world. But our sinnes are often an occasion that God doth not impart his graces so abundantly as he would. Yet I will say one thing, which I holde for truth, that although the first entry of the Gospel hath not beene accompanied (in many places), with such synceritie and Christian meanes as they should have vsed; yet God, of his bounty, hath drawn good from this evill, and hath made the subiection of the Indians a perfect remedie for their salvation. Let vs consider a little what hath beene newly converted in our time to the Christian Religion as well in the East as in the West, and how little suretie and perseverance in the faith and Christian religion there hath beene, in places where the new converted have had full libertie to dispose of themselves, according to their free will. Christianitie, without doubt, augments and increaseth, and brings forth daily more fruite among the Indian slaves: and contrariwise ruin is threatened in other partes where have beene more happy beginnings. And although the beginnings at the West Indies have beene laboursome, yet our Lord hath speedily sent good workemen and his faithfull Ministers, holy men and Apostolical, as Friar Martin de Valencia, of the order of S. Francis,

Friar Domingo de Betanzos, of the order of S. Dominicke, Friar Juan de Roa, of the order of S. Austin, with other servants of our Lord, which have lived holily, and have wrought more then humaine things. Likewise, Prelates and holy Priests, worthy of memory, of whom we heare famous miracles, and the very acts of the Apostles : yea, in our time, we have knowne and conferred with some of this qualitie.

But for that my intention hath beene onely to touch that which concernes the proper history of the Indians themselves, and to come unto the time that the Father of our Lord Iesus Christ saw fit to show the light of his word vnto them ; I will passe no farther, leaving the discourse of the Gospel at the West Indies for another time, and to a better vnderstanding: Beseeching the Soveraign Lord of all, and intreating his servants humbly to pray vnto his Divine Maiestie that it would please him of his bountie often to visit and to augment by the gifts of heaven this new Christendome, which these last ages have planted in the farthest bounds of the earth. Glory, Honour, and Empire be to the King of the ages for ever and ever. Amen.

THE END.

INDEX.

CONTENTS.

	PAGE
I.—GENERAL INDEX	535
II.—DESCRIPTIVE LIST OF NAMES OF PLACES IN PERU	539
III.—QUICHUA WORDS	542
IV.—DESCRIPTIVE LIST OF YNCAS MENTIONED BY ACOSTA	544
V.—YNCA SUCCESSION. THE "AYLLUS" OR LINEAGES	546
VI.—MEXICAN NAMES	547
VII.—DESCRIPTIVE LIST OF SPANIARDS AND OTHER EUROPEANS	548

I.

GENERAL INDEX.

Agatarchides, book on the Erythrean Sea reported by Phocian in his *Bibliotheca*, on refining gold, 192
Alligators, 148
Alligator pear, 250
Ambrose, St., 8, 9
America. (*See* Indies.)
Ancestors, worship of, 312
Antarctic pole (*see* Pole), unknown lands, 170
Antilles soil, products, 169
Antipodes, views of Lactantius and St. Augustine on, 4, 19, 22, 23
Architecture of the Yncas, 415
Arctic zone, extent of land unknown, 171
Aristotle, correct opinion of the shape of the heavens, 4, 9, 21; believed the torrid zone to be uninhabitable, 25, 27, 29, 32, 75, 81, 96; ignorant of the compass, 48; on birds, 275
Armadillos, 283
Astrolabe, height of the sun by the, 15
Atlantis, isle of, 64, 65, 90, 102
Augustine, St., views as to the shape of the heavens, 3; doubt as to the South Pole, 4, 6, 9, 19; denied the Antipodes, 22, 23, 32, 45, 47, 187; on beasts found on islands, 58; on extension of Christianity, 528
Australia, conjectured existence, 170
Avicenna, 91
Axi, 239, 240. (*See* Pepper.)
Aymara dictionary, by Bertonio, v

Bacalaos, 60
Balsam, 257, 258
Bamboos, 263
Baptism, rite of Mexicans resembling, 369
Barter, use of, 189
Basil, St., 8
Batatas, 235
Bears, 274
Bees and honey, 274
Beer. (*See* Chicha, Maize.)
Bezoar stones, 288, 292
Birds in the Indies, 275, 279
Bogos, 151
Bonzes, 339
Brazil wood, 260
Bridges, 416, 417

Cacao, 244
Calendar, Mexican, 392
Calibashes, 238
Camels in Peru, 272
Camotes, 235
Canary Isles known to Pliny, 33; name, 34
Canopus star, 14
Capsicum, 239
Carthage, voyages of ships of, 55. (*See* Hanno.)
Cassava bread, 232, 233
Cassia fistula, 260
Cattle in the Indies, 271
Cayman. (*See* Alligator.)

INDEX.

China, learning, 401 ; writing, 408
Chirimoya, 251
Chocolate, 244, 245
Chicha, 230, 231
Chrysostom, St., notion of the shape of the earth and heavens, 1, 2
Climate, in tropics, 76, 77 ; beyond tropics, 77 ; of Chile, 78 ; dry regions in tropics, 88 ; lofty regions the coldest, 96 ; cause of rainless belt on coast of Peru, 166, 167
Cloth made from llama wool, 289
Coca, 164, 189, 244, 245, 246
Cocoa nuts, 253
Cochineal, 248
Comet in 1577, motion, 122
Compass, ancients ignorant of, 48, 49; virtues of the load-stone, 50, 51 ; variation, 52
Condors, 279
Confession, used in Peru, 361, 362
Conversion, divine arrangements for, 528
Copal, 260
Corn. (*See* Maize.)
Cotton, 249
Council of Lima, vii

Dances, Peruvian and Mexican, 444 to 446
Dantas or tapirs, 283
Datura, sent to Spain by the Viceroy Toledo, 255
Dead, the worship of, 311, 313 ; customs in Peru, 314 ; in Mexico, 315
Deluge, tradition of, 70
Devil, the, his pride the cause of idolatry, 298 ; his malice, 300, 307 ; his cunning, 324 ; monks invented by, 334; penance invented by, 337 ; sacrifices to, 340 ; cruelty of unendurable, 352 ; imitates the sacraments of the church, 354, 356 ; confession to, 360 ; unction, 364 ; illusions of, 371 ; invents a Trinity, 373, 377 ; final defeat of, 381
Dioscorides, 48
Dogs in the Indies, 272
Drugs, 260
Dyes, 260

Earth, shape, opinion of the ancients, 1 ; part discovered, 18 ; circumnavigated, 4 ; round, 5 ; rests upon nothing, 10 ; distribution of land and sea, 17 ; worship of, 304
Earthquakes, 178, 179, 180
Eclipses, proof of roundness of the earth from, 5
Emeralds, 37, 224, 225

Equinoctial, nature of, 73 ; crossed by the author, iii, 90
Eudoxus, voyage of, 33
Eusebius on prognostications, 506, 508

Fathers of the Church (*see* Augustine, Chrysostom, Jerome, Gregory Nazianzen), they may err, 3
Feathers, art of working in, 280
Fig tree at Mala, 268
Fishery. (*See* Pearl.)
Fishing (*see* Whale), in balsas, 150 ; by Chirihuanas, 151 ; in Lake Titicaca, 151
Floating gardens at Mexico, 469
Flocks. (*See* Llamas.)
Florida, strait of, 140
Floripondio, flower, 255. (*See* Datura.)
Flowers in the Indies, 255
Frost-bite, a man lost his toes by, 133
Fruits of the Indies, 236, 237
Fruit trees, 265, 268, 249, 251, 252

Gallinazos or turkey buzzards, 279
Gardens, floating, at Mexico, 469
Genoa, great emerald at, 225
Ginger grown in the Indies, 239
Giants, bones of, found at Manta and Puerto Viejo, 56 ; in Mexico, 454
Gold in the Indies, 190 to 193
Granadilla, fruit of the passion flower, 256
Gregory Nazianzen, 8, 23
Guano on the coast of Peru, 281
Guayavos, fruit, 250
Guinea, New, opinions concerning, 18, 47

Hanno, voyage of, 32
Head-dresses, 422
Heavens, shape of, notion of St. Chrisostom, 1, 2 ; of Theodoret, 2 ; of Lactantius, 2 ; of St. Jerome, 2 ; of Procopius, 2, 8 ; of St. Augustine, 2 ; true shape, 5, 7, 12 ; proof from eclipses, 5
Hispaniola said to be Ophir, 37
History, profit to be derived from, 388, 448
Horses in the Indies, 271
Human sacrifices, 320, 346 to 350
Humming-birds, 279

Idols (*see* Devil), in Mexico, 318, 319, 369 ; in Peru, 371 ; the testimony of, 508
Idolatry, forms of, 303 ; sin of, 306
Imagination, uses of, 20
Indian corn. (*See* Maize.)
Indies (America or New World), by what means men might have first

INDEX. 537

reached, 45, 46, 47; discovered by chance, 54, 56; possibly peopled by land, 57, 455; how beasts reached the New World, 58, 59, 62, 63; idea of Jewish descent of Indians refuted, 67, 68; report of the Indians as to their origin, 70, 71; origin of native civilisation, 72; shape of the Indies, 182; how there can be animals peculiar to, 277. (*See* Mexico, Peru.)
Irrigation, 159
Isaiah, prophecy, 44, 528
Isthmus of Panama, question of a canal, 135
Japan, confession used in, 363, 369
Jerome, St., view as to the shape of the heavens, 2, 8, 15, 32; on Tarshish, 41
Jesuits in Peru, iv; their work, v. (*See* Acosta.)
Josephus on the position of Ophir, 39

Lactantius, view as to the shape of the Heavens, 2; held that there were no antipodes, 19, 32; on the testimony of idols, 508
Lakes in the Andes, 152; in Mexico, 153
Lima, synod of; rule as to Indian marriages, 426; council of, vii
Liquidambar, 259
Llamas, 289; as beasts of burden, 290; diseases, 291
Llanos, 237

Macrobius, 24
Magdalena, river, 158
Maize, 228; harvest, 229; uses, 229, 230; beer made from, 230, 231
Malacca, 33
Manatis, 146
Marriages, in Mexico, 370; in Peru, 369, 424, 425, 426
Mela, Pomponius, 24
Mendocino, cape; nothing known beyond, 18, 60, 171
Menomotapa, climate of, 94
Mercury. (*See* Quicksilver.)
Messengers, Peruvian, 409, 423
Metals, abundance in the New World, 185; gold and silver, 186-9; mineral wealth of Peru, 187; gold, 190 to 194; silver, 194; mines of Potosi, 197; quicksilver, 211
Mexico. (*See* under Mexican Section.)
Milky way, 7, 14, 15
Miracles, of rain at Santa Cruz de la Sierra, 524; worked by Spaniards, 525; in defence of Spaniards, 526; at siege of Cuzco, 526. (*See* Omens.)

Monks, 334, 335; in Mexico, 336
Monkeys, 284, 285
Months, Peruvian, 374, 375
Moon, eclipses of, prove the earth is round, 5; worship of, 304
Mulberries, 269
Mummeries of the Yncas, 432

Nature, study of, 184
Navigation, Portuguese expert in the art of, 15
Nepos, Cornelius. (*See* Pliny.)
New World. (*See* Indies.)
Nicaragua, 127
Night, cause of darkness, 5
Nile, sources unknown to the ancients, 27; cause of inundation, 78
Nobility, Mexican, 438
North-west passage, 18, 141

Obadiah, his prophecy, 43
Olives, 269
Omens, before the Spaniards arrived in Mexico, 506, 510
Ophir, whether Peru is? 37, 38; true position, 39; view of Josephus, 39
Oranges, plant themselves, 265
Orejones, nobles of Peru, 413
Orosius, Paulus; on omens, 507

Panama, climate, 77; sea, 99; tide, 144
Paraguay, inundations of, 78, 158
Pearl, fisheries, 226, 227
Peccaries, 282
Penance, of Mexican priests, 338; of Peruvians, 339
Peru, seasons in, 80; winds, 111, 112; rivers, 158; coast valleys, 160, 161; Sierra, 161; physical features, 164; rainfall, 165; use of rainless coast, 166, 167; mineral wealth (*see* Metals), 187; animals, 273, 282; birds, 275; vicuñas, 287; llamas, 290; maize, 226; roots, 232; pepper, 239; fruit, 251, 252; religion, 301, 302; deities, 304; idols, 308, 371; superstitions, 309; worship of the dead and ancestors, 311, 312, 313; temples, 325, 326; convents of virgins, 331, 332; confession, 361, 362; sacrifices, 340 to 344; sorcerers, 362, 367; marriage, 369; dances and music, 445
Pepinos, 237
Pepper, 239, 240
Phocion. (*See* Agatarchides.)
Picture writing, 403
Pilot fish, or Romeros, 147
Pine apples, 236

Pitch, springs of, 165
Planets, motions of, 7
Plantains, 241
Plants, introduced from Spain, 265
Plate, river, inundations of, 78, 158
Plato, his opinion touching the New World, 36; on Atlantis, 64, 65, 90
Pliny, held the opinion of Aristotle as to the tropics, 29, 32; ignorant of the compass, 49, 55; mentions crocodiles, 148; on emeralds, 225; silver, 201; pearls, 227; birds, 275; millet, 231; plane, 241; on a story, in Cornelius Nepos, of Indians coming to the King of Suevia, 55; death of, at Mount Vesuvius, 177; on mines in Spain, 201; on quicksilver, 213
Poles, arctic, 171; antarctic, 16, 28, 170
Portuguese, expert in navigation, 15
Potatoes, 233
Prickly pear, 463
Priests, in Mexico, 330; training of, in Mexico, 443
Ptolemy, believed the tropics to be habitable, 91
Puna of Peru; intense cold, 132, 133
Purgatives, 261

Quicksilver, properties of, 211; discovery of in Peru, 214, 215; method of preserving workmen from poison of, 212

Rainbow, worship of, 304
Rainfall, in the tropics, 79; effects of sun on, 84; tempers heat, 91; rain bearing winds, 127; in Peru, 165; cause of no rain on the Peruvian coast, 166
Rice, 234
Rivers, Amazons, 156; *Pongo*, or rapid, 157; of Peru, 158
Romero. (*See* Pilot fish.)
Roots, edible. (*See* Potatoe, 233; Oca, 235; Camote, 235; Yuca, 233, 235.)

Sacraments of the Church, counterfeited by the Devil, 346, 354
Sacrifices—human, in Mexico, 323; 346 to 350; Peruvian, 340, 341, 342, 343, 344
Salomon Isles; opinions as to position, 18; discovery of, 46, 47, 115
Salt, fountain of, 155
Saltpetre, cools water, 95
Sarsaparilla, 156
Schinus Molle, 264

Schools, in the Mexican temples, 442
Sea sickness, 129
Seneca, thought to have alluded to the West Indies, 34, 35
Sharks, voracity of, 147
Sheep, in the Indies, 270
Sickness, at sea, 129
——— at great heights, 130, 131
Silver, in the Indies, 194; Pliny on, 201; refining, 217; engines for grinding ores, 222; trial of, 223
Sloths, 284
Snow blindness, 288
Sorcerers, 362, 367, 498
South sea, 56, 134
Southern cross. (*See* Stars.)
Springs, hot and cold, 154, 156; of pitch, 155; of salt, 155; at Guayaquil, flowing by sarsaparilla, 156; rising on Vilcañota, 156
Stars, their motions, 6; Southern Cross, 14; and Canopus, 64; milky way, 7, 14, 15; in southern hemisphere, 14; names in Peru, 305
Storax, 260
Strabo, on balsam, 258
Suevia, King of. (*See* Pliny.)
Sugarcane, 269
Sumatra, 49, 55, 91
Sun, effect on rainfall, 78, 85; on vapours, 86; worship of, 303, 304, 305; argument against its being God, 310
Synod of Lima; rule as to marriages, 426

Tapirs, 283
Tarshish, 38, 40, 41, 42
Tarugas, 288
Temples, in the Indies, 325
Theodoret; opinion as to the shape of the Heavens, 2; on the position of Tarshish, 41
Theophilus, 2
Theophrastus, 48; emeralds mentioned by, 225
Thunder and lightning, worship of, 304
Tides, 143, 144, 145
Timber trees, 262
Time, change of, in sailing round the world, 173
Tobacco, 261
Totora, 235, 417
Trinity of the Peruvians, 373
Tropics, held to be uninhabitable, 25; climate of, 76, 77; rainfall, 79; abound in water and pastures, 81; dry regions of, 88, 89; moderate heat in, 90, 91, 94, 95; length of

INDEX. 539

days and nights, 92; cold winds, 98; pleasant life in, 101
Tunal (prickly pear), 463

Unction used in Mexico, 364

Vermillion, 214, 216
Victoria, ship which has encompassed the earth, 4
Vineyards, 267, 268; in Peru, 168
Virgins, convent of, in Peru, 232; in Mexico, 333

Warfare, Mexican, 440
Whale fishing, 149
Winds, cause of temperate climate in the tropics, 98; land and sea breezes, 100, 126; their properties and causes, 105; in Peru, 111; trade winds, 113, 115; names of winds, 118, 119; cause of trade winds, 121; cause of westerly winds outside the tropics, 124; rain-bearing winds, 127

Yguanas, 283

Zarephath, supposed to be Spain, 43
Zones. (*See* tropics), 25; southern, 28; burning, 72 (*see* equinoctial); held to be uninhabitable, 74; burning zone very moist, 75 (*See* Arctic, Antarctic).

II.

NAMES OF PLACES IN PERU MENTIONED BY ACOSTA.

(G *denotes places also mentioned by Garcilasso de la Vega.*)

Acoria, 216. Village, a native of which, named Nauincopa, discovered a quicksilver mine in Huancavelica. Acoria is now a district in the department of Huancavelica, with a small village of 646 inhabitants.
Amazons, great river of, 82.
Andahuaylas, 165, 430. A town, capital of the province of the same name, on the road from Ayacucho to Cuzco; in 13° 36' 54" S. lat. It is situated in a long fertile valley, enjoying a temperate climate, and surrounded by mountains. G.
Angoango, 180 (Ancu-ancu). A hamlet in the parish of Achacache, on the east side of lake Titicaca.
Anti-suyu, 414. The eastern division of the Empire of the Yncas. G.
Apurimac, 151. A great river which, with its tributaries, drains the mountainous country round Cuzco, and eventually falls into the Ucayali. G.
Araucanos, 170, 410, 427, 530. The independent Indians in the south of Chile. G.
Arena, 168. A mountain near Lima
Arequipa, 151, 161, 166, 167, 173. Capital of the department of the same name, in 16° 24' 28" S. lat., in a fertile valley at the foot of the volcano of Misti. Arequipa was founded by order of Pizarro, in 1540. G.
Arica, 56, 218. A seaport of Peru.
It has been several times destroyed by earthquakes. G.
Atico, 167. On the coast, between Yca and Arequipa. Occasional rain there. G.
Callao, 95. The port of Lima, in 12° 4' 15" S. lat.
Cañaris, 428, 532. A powerful tribe in the kingdom of Quito. G.
Cañete, 150. A town on the coast, south of Lima, in a plain covered with sugar cane. It was founded by the Viceroy Marquis of Cañete. G.
Capachica, 290. The weavers of *ccompi* lived in this province, on the shores of lake Titicaca. The promontory of Capachica forms a bay in the north-west end of the lake, 15° 44' 28" S. lat.
Caravaya, 39, 192. A province of the department of Puno, on the eastern side of the Andes. Its forests are watered by streams famous for their gold washings. G.
Cavanas, 131. Corruption of Cahuana. Several places of this name. One near Huamachuco, another in Ancachs, another near Lucanas, another in the department of Puno.
Caxamarca, 432, 434, 435. Corruption of *ccasa*, ice; and *marca*, a town. In a large plain, at the foot of the eastern Andes, in 7° 9' 31" S. lat. Here the Ynca Atahualpa was arrested, and put to death by Pizarro. G.

INDEX.

Chachapoyas, 163, 180. A province and town in the department of Amazonas, in 6° 7' 41" S. lat. G.

Chancas, 431. A warlike tribe of the Ynca nation, round Guamanga, and extending as far as the Apurimac. G.

Charcas, iv, 150, 155, 274, 525. A great province of the old Vice-royalty of Peru; the modern Bolivia. G.

Chichas, 417. A tribe in the southern part of Upper Peru (modern Bolivia). G.

Chincha-suyu, 414. The northern division of the Empire of the Yncas. G.

Chirihuanos, iv, 72, 150, 530. A warlike tribe in the forests to the east of the Andes, in Upper Peru (modern Bolivia). G.

Chucuito, 161, 362; lake, 416. A town on the western shore of lake Titicaca. The lake itself was sometimes called "of Chucuito". Lat. 15° 54' 10" S., about 12,000 feet above the sea. G.

Chumbivilicas, 198, 199, 417. The dancers of the Ynca court. Their province is near Cuzco, in the valley of the Apurimac. G.

Chunchos, 427, 530. Wild Indians in the forests east of the Andes. G.

Chuqui-apu (see La Paz). From *chuqui*, a lance in Quichua, or gold in Aymara; and *apu*, chief. See *G. de la Vega*, i, p. 225. On this site the city of La Paz was founded. G.

Colla-suyu, 361, 414. The southern division of the empire of the Yncas. G.

Collao, 83, 95, 151, 155, 361, 416. The region comprised in the northern half of the basin of lake Titicaca.

Collahuas, 131. In the province of Huaras, north of Lima, a pass over the Andes. Another of the same name near Arequipa. G.

Coaillo, 368. A province where there were many witches.

Cunti-suyu, 414. The western division of the empire of the Yncas. G.

Cuzco, 155, and *passim*. The capital of the empire of the Yncas. G.

——— Miracle at the siege of, 526.

——— Hanan, 71, 429. (Upper). G.

——— Urin, 71, 429, 436. (Lower). G.

Desaguadero, 416. The river which drains lake Titicaca, flowing southwards. G.

Guamanga, 216 (correctly Huamanca), now called Ayacucho.

Founded by Pizarro, 9 Feb. 1539. Lat. 13° 8' 45" S. G.

Guayaquil, 156. The sea port of Quito. G.

Huanca, 199. A tribe of the Ynca nation in the valley of Xauxa. G.

Huancavelica, 154, 160, 215 (correctly Huanca-villca), in 12° 48' 38" S. lat. Capital of the department of the same name, in the cordilleras, once famous for its quicksilver mines. G.

Huarco, 150. The plain on the coast, now known by the name of Cañete. G.

Huarochiri, 368. Folk-lore of, v. A province of the department of Lima, in the maritime cordilleras: between 11° 20' S., and 12° 35' S. It contains the sources of the coast rivers, Rimac, Lurin, and Mala.

Juli, station of the Jesuits at, v. On the banks of lake Titicaca.

La Paz, 180. A town to the south of lake Titicaca, now the commercial capital of Bolivia. Founded in 1548 by Alonzo de Mendoza, by order of the President Gasca. Lat. 17° 30' S. The bishopric of La Paz dates from 1605

Lima, 46, 111, 127, 426, 432. The capital of Peru. Founded by Pizarro, January 18, 1535, in 12° 2' 34" S. Called also the City of the Kings.

Lucanas, 131, 230, 417. Bearers of the Ynca's letter. A province in the department of Ayacucho, properly Rucanas. G.

Mala. A valley on the coast of Peru, south of Lima. Fig-tree in, 268

Manchay, 368. The *lomas*, near Lurin, on the coast, are so called; also an hacienda near Pachacamac.

Manta, 225. On the sea-coast of the kingdom of Quito. G.

Marañon, 82, 83. The upper course of the great river Amazon. G.

Nasca, 308. A town and valley on the coast, yielding vines and cotton, and irrigated by ancient channels. G. Correctly Nanasca.

Ollantay-tambo. (*See* Tambo.) Ynca ruins. G. de la Vega calls it simply Tampu. G. In the valley of the Vilcamayu, near Cuzco.

Omasuyo, 151, 429. A province on the eastern shores of lake Titicaca. Correctly Uma-suyu. G.

INDEX. 541

Paccari-tampu, 71. A place in the province of Paruro, near Cuzco. Several traditions point to this place as the cradle of the Ynca race. It is said that Manco Ccapac first appeared here. G.

Paria lake, 151, 283. In the south of Bolivia. The river Desaguadero, draining the lake of Titicaca, empties its waters into the salt lake of Paria or Aullagas. G.

Pariacaca, 131. A pass over the maritime cordillera of the Andes, in the province of Huarochiri.

Pasto, 427. The most northern province of the kingdom of Quito, but now in Colombia. G.

Patallacta, 432. An estate in the province of Paucartambo, near Cuzco. There is another place of the same name in Tayacaja, a province of Huancavelica.

Paullo, 429. An estate or farm, near Calca, in the valley of the Vilcamayu (department of Cuzco).

Payta, 147. A seaport in the north of Peru, in 5° 6' S.

Paytiti, 82, 156, 171. A fabulous kingdom in the forests east of the Andes.

Pilcoçones, 427, 530

Popayan, 95. A town north of Quito, in Colombia : in the province of Cauca.

Potosi, 90, 152, 196, 197, 198, 199, 200, 203, 218, 222, 525. A famous silver yielding district and town in Upper Peru (now Bolivia), in the province of Porco. Correctly Potocchi. G.

Puerto Viejo, 225. A seaport on the coast of the kingdom of Quito. G.

Porco, 196, 199, 200, 201. A province in Upper Peru, in the centre of which is Potosi.

Quito, 90, 175, 433. Capital of the kingdom of the same name, nearly on the equator ; the most northern part of the empire of the Yncas. G.

Rucana. (*See* Lucanas.)

Runahuanac, 281. Corruptly Lunahuana ; in the province of Cañete, south of Lima. The town is on the left bank of the river Cañete. G.

Salinas, 192

San Blas parish, in Cuzco, 432

Saruma, 192. Mines in the Government of Salinas.

Soras, 131. A district in the province of Lucanas, department of Ayacucho. G.

Sta. Cruz de la Sierra, 170, 189, 524 A town and province in the eastern part of Bolivia

Tambo, 415. The great ruins of Ollontay-tambo in the valley of the Vilcamayu. G.

Tiahuanaco, 71, 415. The great ruins near the south shore of lake Titicaca. G.

Titicaca, iv, 71, 83, 151, 165. The great lake. The boundary between Peru and Bolivia passes across it. It is 40 leagues long by 20 broad, between 15° 59' 57", and 16° 3' 40" S. lat. ; 12,545 feet above the level of the sea. G.

Tanaca ñuñu, 232

Tarapaya, 153, 218, 222. Near Potosi. An extensive and fertile plain

Toto-cache, 432 (correctly Toco-cachi), a suburb of Cuzco, now the parish of San Blas. G.

Truxillo, 167. City founded by Pizarro in 1535. The bishopric erected in, 1609. In 8° 6' 9" S. lat., near the shores of the Pacific. G.

Ttahuantin-suyu, 414. "The Four Provinces". The general name for the empire of the Yncas. G.

Tucapel in Chile, 410, 427, 530

Tucuman, 274. A province south of Charcas, originally in the Viceroyalty of Peru, afterwards in that of Buenos Ayres. G.

Tumbez, 61. The most northern port in Peru, where Pizarro landed in 1526. G.

Tumipampa, 432. A province in the south of the kingdom of Quito. G.

Uros, 83. A tribe of Indians living among the reed beds in the southwest of the lake of Titicaca. G.

Valdivia, 192. A town in the south of Chile.

Vilcabamba, 435. There are several places in Peru called Vilcabamba. The district of Vilcabamba, to which the Yncas retired, is a mountainous tract north of Cuzco, bordering on the forests east of the Andes.

Vilcañota, 156. A snowy peak on the eastern cordillera, in 14° 28' 30" S. lat. ; 17,000 feet above the sea. It means "the House of the Sun" in the Colla language. *Vilca*, the sun ; and *ñuta*, a house. G.

Xauxa, 165, 272, 416. A town in the fertile valley of the same name, in 11° 49' 38" S. lat., between the maritime and eastern cordilleras of the Andes ; properly Sausa. G.

542 INDEX.

Yca, 56, 150. A province on the coast of Peru, yielding cotton and wine. The town is in 14° 4' 33" S. lat.
Yscaycingas, 427

Yucay, 155, 165. A village, where there were Ynca palaces and baths in the valley of the Vilcamayu (also called, in this part, the valley of Yucay), near Cuzco. G.

III.

QUICHUA WORDS IN ACOSTA.

Acca, 230. Fermented liquor or Chica. See *G. de la Vega*, i, p. 298
Aclla, 232. Chosen. *Aclla-cuna*, Virgins of the Sun. *G. de la Vega*, i, 292 ; ii, 250
Alco, a dog (canis Ingæ), 272
Alpaca, 277, 341
Amaru, 435. A serpent. See *G. de la Vega*, ii, p. 352
Anas, 59. A small fox.
Apachita, 308, 309. Apachecta, the dative of the present participle of *Apachini*, I carry. See *G. de la Vega*, i, 117. *Muchani*, I worship. *Apachecta muchani*, "I offer up thanks by throwing a stone on a heap by the road side", on the summit of a pass. Two words used by the Indians on reaching the top of a pass.
Apu, 373. Chief.
Apu-panaca, 332. Officer in charge of a convent. See also *Ramos*, cap. 9, and *Ondegardo*, p. 165
Arepas, 230
Auasca, 434, 435. Coarse cloth.
Atahualpa, 434, 435. For the derivation see *G. de la Vega*, i, lib. ii, cap. 23
Ayamarca, 376. Month of October.
Ayllu, 429, 432. Lineage. See *G. de la Vega*, i, 67
Ayma, 377. A song. See *Molina*, p. 89
Aymuray, 373. April and May. Time of harvest. Ayrihuay, *Molina*, 33, 52
Aucaycuzqui Ynti-raymi, 374. June
Cachi, 432. Salt (in Toco-cachi).
Camac (from *Camani*, I create). In the word *Pachacamac*, which see.
Camay, 373. December.
Carachi, a disease in llamas, 291, 420. See *G. de la Vega*, ii, 378
Catuchillay, 304. A star worshipped by shepherds, near the milky way.
Catuilla, 304. A name for thunder.

Cavi, 235. An edible root.
Cayo, 375. Dancing. See *Molina*, p. 89. A playing on drums and singing.
Ccapac, rich. 420, 433
Ccapac Raymi, 354 (see Raymi).
Ccompi, 289, 340, 412, 417. Fine cloth. See *G. de la Vega*, ii, 324
Ccoya, 411. Queen.
Ccoya Raymi, 355. Tenth month
Chacana, a star, 305. Also *Balboa*, 58
Chacra, 374. A farm.
Chacu, 151, 273, 287. A hut. See also *G. de la Vega*, ii, 109, 115
Chahua huarqui, the eighth month, 375
Chaquira. Minute beads. *Cieza de Leon*, cap. xlvi. Also *G. de la Vega*, ii, 338
Charqui, 289. Dried meat ; whence jerked meat.
Chasqui, 409, 423. A messenger.
Chicho, 526. Misprint for Ychu.
Chinchilla, 283
Chirimoya, 251
Chunquinchincay, 305. A star.
Chuñu. Frozen potato, 165, 233. *G. de la Vega*, ii, 17, 359
Chuquilla. A name for thunder, 304, 341, 373. *Chuqui*, a lance. *Yllani*, I shine. *Yllapa*, a thunder bolt.
Churi, 373. Son.
Coca, 164, 189. Account of, 244, 245, 246
Cochuchu, 235. An edible root.
Collca, 304. The Pleiades. See also *G. de la Vega*, ii, 237, and *Balboa*.
Contesisca, 342. A sacrifice.
Cuntur, 279. Condor.
Curaca, 375. A chief
Cusi, 434. Joy.
Cutec, from *cutini*, I overturn. See Pachacutec
Cuy, 283, 340. A guinea pig. *G. de la Vega*, ii, 118, 233, 384
Guaras, 373. *See* Huaras.
Hanan, 71. Upper.
Hatun, 373. Great.
Hatuncuzqui, 373. May.

INDEX.

Homaraymi Punchaiquis,376. Eleventh month.
Huaca, 300, 308, 318, 323, 325, 340, 355, 361, 373, 375, 412. Sacred.
Huaccha, 420
Huallavicsa, 349. Sacrifice.
Hualpa, 276, 434. A fowl (in Atahualpa.
Huaman, 436. Falcon (in the name Tarco-huaman).
Huanacu, 277, 341
Huanani, 281. I warn—In the name. Lunahuana (Runahuanac).
Huanu, 281. Guano. See also *G. de la Vega*, ii, p. 181
Huara (Guaras) 373. Breeches.
Huascar, 434. A chain.
Huasi, 332. A house.
Huauque, 312, 323, 373. Brother. *G. de la Vega*, i, p. 314
Huayna, 198, 313, 433. Youth.
Huayra, 195, 196, 209, 210. Wind. Air.
Hunu, 414. An officer over 10,000.
Inti raymi (*see* Ynti).
Ituraymi, 376. (Ytu).
Llallahuas, 309. A kind of potato.
Llama, 288 *et seq.* 420
Llimpi, 215. A purple colour. See *G. de la Vega*, ii, p. 473
Lloque, 355, 436. Left-handed.
Locro, 234. A kind of potato.
Machachuay, 305. Serpent. A constellation. *G. de la Vega*, ii, 240, 385
Mama, mother (in mama-cocha, Pucha mama, etc.)
Mama-cocha, 303. The sea. *G. de la Vega*, i, p. 293, 300, 302
Mama-cuna, 332, 355. Matrons of the Virgins. *G. de la Vega*, i, 293, 300, 302
Mamana, a constellation, 305. *Balboa*, p. 58
Mani, 235. An edible root.
Miquiquiray, 305. A constellation, *Balboa*, p. 58.
Mirco, 305. The Southern Cross. *G. de la Vega*, ii, p. 476, and *Balboa*, p. 58
Mitimaes, 413. Emigrants. *G. de la Vega*, ii, p. 476
Morochi, 229. A kind of maize. *G. de la Vega* has *Muruchu*, ii, p. 355
Mulli, 264. The molle tree (*Schinus Molle*). *Cieza de Leon*, chap. cxii. See also *G. de la Vega*, i, p. 187 ; ii, p. 364, 367
Mullu, 340. A shell.
Mutti, 230. Boiled maize. *G. de la Vega*, ii, p. 357

Oca, 235 (*Oxalis-tuberosa*). An edible root
Opa-cuna, 362, 369. Baths. Correctly *Upa* from *Upani*, I wash.
Otojo (*see* Usuta)
Otoronco (*see* Uturuncu).
Paccari, 71. Morning. Paccari-tampu and its legend are mentioned by *G. de la Vega*, i, lib. i, cap. 15, 18. *Fernandez*, Pt. ii, lib. iii, cap. 5, p. 125. *Balboa*, *Ondegardo*.
Pacha. Earth. (in the words Pachacamac, Pachayachachic, etc.)
Pachacamac, 301, 325, 327 ; " Creator of the World", *G de la Vega*, i, p. 106 ; and ii, pp. 9, 38, 58
Pachacutec, the Ynca, 430.
Pacha-mama, 304. "Mother earth".
Pacha yachachic,301,418,434."Teacher of the World". See *G. de la Vega*, i, p. 109 ; ii, p. 56
Palta, 250
Panaca. See *Apu-Panaca*.
Pancuncu, 376. A torch. See *G. de la Vega*, ii, p. 232
Papa, 165, 235, 236, 308. Potatoe. *G. de la Vega*, ii, p. 517, 213, 359
Pirua, 374. A granary.
Pucara, 427. A fort.
Puclla, 444. A sham fight. Warlike exercise. The word occurs in one of the prayers given by *Molina*, p. 31. From *Pucllani*, I play.
Punchau, 326. Day ; Idol of the Sun. See *G. de la Vega*, i, p. 182
Puncu (pongo), 156. Door. *G. de la Vega*, ii, p. 240, 312
Pururaucas, 432. Certain Idols. See *G. de la Vega*, ii, p. 57
Quinua, 198. (*Chenopodium Quinoa*), *G. de la Vega*, ii, 5, 7, 213, 357, 367
Quipu, 406, 407, 426
Quipucamayoc, 71, 72, 406, 415. See *G. de la Vega*, ii, p. 123
Quirau, 429. A cradle. (In Vicaquirau.)
Quiso, 342. An assembly of birds for sacrifice.
Raymi, 354, 372. Festival.
Raymi cantara rayquis, 376. Festival.
Runa, 281. Man. *G. de la Vega*, i, p. 35; ii, 181
Runtu, 276. Egg. *G. de la Vega*, ii, p. 89, 481
Saparisca, 342. Sacrifice.
Sapay, 301. Sole. Only. *G. de la Vega*, i, p. 95, 324
Situa, 355, 375. Festival.
Sora, 230. A strong liquor. *G. de la Vega*, i, p. 277

Sucanca, 395. Solstitial pillars at Cuzco.
Suchi, 151. Fish in lake Titicaca. *G. de la Vega*, ii, p. 402
Suyu, 361. Province.
Tampu, 287. Inn.
Taqui, 445. Music.
Tanga-tanga, 373. Idol at Chuquisaca. See *G. de la Vega*, i, p. 120. Represented the Trinity.
Tarco-huaman (*See* Huaman).
Ttahuantin - Suyu, 414. The four provinces. The empire.
Tiçi Viracocha, 307. Perhaps Tiçi, from *Atic*, conquering. See Quichua prayers, given by Molina.
Titu, 38, 434. A proper name.
Toco, 432. Window (in Toco-cachi).
Tomahaui, 197. A cold wind.
Topatorca, 305. A star.
Ttanta, 228, 236. Bread. *G. de la Vega*, ii, p. 357
Uchu, 237. Axi pepper.
Uiscacha, 283 (*Lagidium Peruvianum*). *G. de la Vega*, ii, p. 384
Upa. *See* Opa.
Urcu, 341. Male.
Urin, 436. Lower.
Urcuchillay, 303. The star Vega. *Balboa*, p. 58
Usachun, 341. From Usachuni, I accomplish.
Usapa, 301. (*See* Sapay)
Usuta, 67, 424. Shoes. See *G. de la Vega*, i, p. 82 ; ii, p. 171.
Uturuncu, 274. Jaquar. *G. de la Vega*, ii, p. 385
Vicuña, 132, 286. See *G. de la Vega*, ii, 117, 378, 383, 384
Vilcaronca, 341. A sacrifice.

Vilca. (In Vilcaronca), 341. Sacred. See *G. de la Vega*, ii, 255, 416. *Molina*, 63, 93, 107
Villca, 368. A tree, the fruit of which is a purgative. (*Mossi*.) The juice is mingled with Chicha.
Viracocha, 301, 304, 307, 418, 428, 434. *G. de la Vega*, ii, 66
Xiquimas, 235. An edible root?
Yachachic. In Pachayachachic, 301, 418, 428. *G. de la Vega*, i, 110. From Yachami, I teach.
Yanlli, 342. A thorny tree.
Yana. Black.
Yanacauri.
Yana-cunas, 368, 433. Indians bound to service. See *Balboa*, p. 120, for the origin of this servitude. See also *G. de la Vega*, ii, p. 411
Yana-oca, 235. An edible root. Black Oca.
Yapaquis.
Ychu, 218, 526. (*Stipa Ychu*). *G. de la Vega*, i, p. 254. (*See* Chicho)
Ychuri, 361. Confession.
Yllapa, 302, 304, 432. Thunder and lightning. *G. de la Vega*, i, 105, 182, 275
Ynca (*passim*)
Ynti, 302, 373. Sun.
—— Apu Ynti, chief sun.
—— Churi Ynti, son.
—— Ynti Huauque, brother.
Yntip Raymi, 374
Ytu, 376. Feast.
Yuca, 232 (*Jatropha Manihot*). But the proper Quichua word is Asipa, or Rumu.
Yupanqui, 355, 356, 411. Virtuous.

IV.

INDEX OF THE YNCAS MENTIONED BY ACOSTA.

Amaru, *see* Tupac Amaru
Atahualpa, 313, 325, 425, 434, 529. Son of the great Ynca Huayna Ccapac, by a Princess of Quito. He usurped the throne of the Yncas from his legitimate brother Huascar. For an account of the sanguinary War of Succession, see *G. de la Vega*, ii, p. 505 to 529. See also *Velasco, Historia de Quito*, vol. ii. *Balboa* also gives a detailed account of the war, which he received from the officers of Atahualpa at Quito. The most authentic account of the arrest of Atahualpa at Coxamarca, and of his judicial murder, is in the narrative of Xeres, Pizarro's secretary. See also my note at p. 102 of my translation of Xeres.

Caritopa, 432. Don Felipe, grandchild of Tupac Ynca Yupanqui.

Ccapac Yupanqui, 436. The fifth Ynca. His reign and death will be found described in *G. de la Vega*, i, p. 234

and 269. His lineage, called *Apu Mayta*, at ii, p. 531. See also *Molina*, 85, 88

Coya Cusilimay, 425. Daughter of Tupac Ynca Yupanqui, and sister of Huayna Ccapac.

Chilicuchi, 434. Atahualpa's general, who took Huascar prisoner. This is Acosta's form of Challcuchima. See *G. de la Vega*, ii, p. 509. Xeres has *Chilicuchima*, p. 84 to 89.

Guaynacapa (*see* Huayna Ccapac).

Huascar, 425, 434, 529. The legitimate son and successor of Huayna Ccapac. See the account of the birth, and of the rope of gold (Huasca) made to celebrate it, in *G. de la Vega*, ii, p. 103 and 422. His accession and war with Atahualpa, ii, p. 505, *et seq*.

Huayna Ccapac, 313, 425, 433, 529. The twelfth Ynca. His name means "the rich youth", or one who, from childhood, has been rich in magnanimous deeds. See *G. de la Vega*, ii, p. 345. His three marriages are given at, ii, p. 351; his conquests, ii, p. 423 to p. 444; his remarkable saying touching the Sun, ii, p. 445; the division of the Empire between his sons, ii, p. 450; his will and death, ii, p. 461; the discovery of his mummified body, i, p. 273; his lineage, ii, p. 532

Mama Ocllo, 425, 434. The mother of Huayna Ccapac. See *G. de la Vega*, ii, 353.

Manco Ccapac, 71, 429. The first Ynca. Acosta says that, after the deluge, he came out of the cave at Paccari-tampu. (*See* Paccari-tampu, in the index of Quichua words.) The various accounts of his origin are given by *G. de la Vega*, i, p. 63 to 85, and *Molina*, p. 6 and 74.

Manco Ynca, 435, 436. Son of Huayna Ccapac. See *G. de la Vega*, ii, pp. 352, 526. He made an heroic resistance against the Spaniards, and besieged Hernando Pizarro in Cuzco, in February 1536. See the second part of *G. de la Vega*, lib. ii, and *Herrara Dec.* v, lib. viii, cap. 4. Manco was murdered by a party of fugitive Spaniards, who had fled to him for refuge. *G. de la Vega*, pt. ii, lib. iv, cap. 7. He left three sons.

Mayta Ccapac, 436. The fourth Ynca. For his reign and conquests, see *G.*

de la Vega, i, p. 173, 210, 233. For his lineage, ii, p. 531.

Pachacutec, 430. The ninth Ynca. The story related by Acosta, respecting his accession, should be told of his father, Viracocha. *G. de la Vega* describes his reign, ii, 201 to 205, and gives his wise sayings, ii, 208, 247.

Paullu Ynca, 435. A son of Huayna Ccapac who was baptized, and accompanied Amalgro on his Chilian expedition. Acosta knew his son Don Carlos. Paullu was personally known to Cieza de Leon. See *Cieza de Leon*, p. 77 and 224. His son Carlos was a schoolfellow of G. de la Vega at Cuzco. His grandson, Melchor Carlos Ynca, went to Spain in 1602, and became a knight of Santiago. See *G. de la Vega*, ii, p. 110, 530. *Balboa*, p. 304.

Quizquiz, 434. A general of Atahualpa. See *G. de la Vega*, ii, p. 484.

Sayri Tupac, 435. Son of Ynca Manco, and grandson of Huayna Ccapac. He was baptized in 1553, and died at Yucay in 1560, leaving a daughter named Ccoya Beatriz, the wife of Don Martin Garcia Loyola. Their daughter was Marchioness of Oropesa. There is a picture of the marriage in the cathedral at Cuzco.

Sinchi Rocca, the second Ynca, is mentioned by Acosta, 436

Tambo, Don Juan, 436

Tarco-huaman, 436. An Ynca not given by other authors. Acosta makes him the son of Mayta Ccapac.

Titu; treasure of Tupac Ynca Yupanqui in power of, 433

Tupac Amaru. Acosta omits the first name, 435. He was the younger son of Manco Ynca, and was unjustly beheaded at Cuzco by the Viceroy Toledo in 1571, iv. See *G. de la Vega*, ii, pp. 264, 273

Tupac Ynca Yupanqui, 425, 433. Father of Huayna Ccapac. Eleventh Ynca. See *G. de la Vega*, ii, 91, 246, 304, 321, 344, 352. His lineage, ii, 531. Discovery of his mummified body, ii, 273

Viracocaha Ynca, 300, 307, 361, 418, 428, 429, 431. The eighth Ynca. His history is given fully by *G. de la Vega*, i, p. 341, and ii, 50, 65,

245, 483, 450. His sayings, ii, 94. His fondness for the vale of Yucay, ii, 87. Discovery of his body by Polo de Ondegardo, i, 273, and ii, p. 91. See also *Cieza de Leon*, p. 226, 308, 332, 338, 355, 363, and *Molina*, p. 12, 90, 92, 95. His lineage was called Socso Panaca. *G. de la Vega*, ii, p. 331. Acosta has Coco Panaca.

Yahuar - huaccac. Acosta spells it Yaguarguaque, 429. The seventh Ynca. See *G. de la Vega*, i, p. 327, 347, 349. He was dethroned for incapacity, ii, 62, 63. His lineage was called Ayllu-panaca, ii, 531. Acosta spells it Ayllu-anaca.

Ynca Rocca, 313, 429. Sixth Ynca. See *G. de la Vega*, i, p. 226, 322, 327, 332. His schools, i, p. 335. Also ii, 247, 248, 354. His lineage, 531

Yncas : their origin, 71, 428 ; use of gold by, 191 ; use of coca by, 246 ; their argument against the Sun being God, 311 ; only confessed to the Sun, 361 ; feasts, 372; divisions of their empire, 414; their edifices, 415; bridges, 416 ; revenues, 418, 419 ; arts, 421 ; head-dress, 422 ; marriages, 424, 425 ; lineage, 429 ; traditions, 430, 431, 432 ; extent of their empire, 427 ; last Yncas, 435 ; succession, 436

V.

YNCA SUCCESSION AND THE *AYLLUS* OR LINEAGES OF EACH YNCA.

HANAN CUZCO.

Yncas.	Lineage (Acosta).	Page of Acosta.	Lineage (G. de la Vega).	No. of souls in 1570.
[1]6.—YNCA ROCCA	Vica-quirao	429	(Vica-quirau)	50
7.—YAHUAR HUACCAC	Ayllu-panaca	429	(Ayllu-panaca)	51
8.—VIRACOCHA	Coco-panaca	429	(Socso-panaca)	79
9, 10. { PACHACUTEC, or YNCA YUPANQUI }	Ynaca-panaca	432	(Ynca-panaca)	99
11.—TUPAC YNCA YUPANQUI	Ccapac sylla	433	(Ccapac ayllu)	18
12.—HUAYNA CCAPAC	Tumi-pampa	433	(Tumipampa)	22

URIN CUZCO.

1.—MANCO CCAPAC			436 (Chima-panaca)	40
2.—SINCHI ROCCA			436 (Raurava-panaca)	74
4.—CCAPAC YUPANQUI			436 (Apu Mayta)	53
3.—LLOQUE YUPANQUI			436 (Hahuarina-panaca)	73
5.—MAYTA CCAPAC			336 (Usca Mayta)	35
TARCO HUAMAN			436	
DON JUAN TAMBO			436	

Descendants in the time of Garcilasso de la Vega 594

---

[1] These numbers show the succession, in one line, according to Garcillasso de la Vega. Acosta makes two lines spring from Manco Ccapac.

INDEX. 547

VI.

MEXICAN NAMES IN ACOSTA.

Acamapich, 436
Acamapixtli, 468, 470. First king of Mexico.
Acatzuitillan, 462
Acopilco, 460
Atlacuyavaya, 460
Axayaca, seventh king of Mexico, 493, 494
Autzol, eighth king, 497
Azcapuzalco, 482
Chalcas, 460, 517
Chalco, 489, 517
Chapultepec, 357, 459, 473. A charming retreat near Mexico.
Chichemecas, 449, 453, 454
Chimalpopoca, third king of Mexico, 472, 473, 475
Cholula, 321, 508, 517. A town in 19° 4′ N., twenty leagues east of Mexico.
Coatepec, 459, in the district of Jalapa; but there are several places of this name.
Copil, 459, 463
Cuitlavaca, 486
Culhuas, peopled Tezcuco, 452, 460
Culhuacan, king of, 461, 466, 476
Cuyoacan, 353, 357, near Mexico. Here Hernan Cortes founded a convent of Nuns, and here, according to his will, he desired to be buried, 483. Sorcerer at, 498
Guatemala, 497
Guatulco, viii, 400. A port on the west coast of Mexico, at the western end of the Gulf of Tehuantepec, in the Oaxaca province. Here Sir Francis Drake landed, viii, n.
Izcoatl, king of Mexico, 371, 436, 476, 482, 485
Iztapalapa, 462
Iztacal, 462
Malinalco, 458, 459. A district in Mechoacan.
Marina, Indian woman. Guide to the Spaniards, 514
Mechoacan, 457, 465, 504. A province on the shores of the Pacific, for eighty leagues.
Mexi, 457
Mexico, lakes, 153, 154; deities, 305; burial customs, 315, 316; idols, 318, 319; gods, 321; sacrifices, 323; temples, 327; priests, 330; virgins, 333; monks, 336; human sacrifices, 323 to 350; festivals, 356, 357, 377 to 384; unction, 364; baptism, 369; marriage, 370; writings, 402; picture writings, 403; records, 404; succession, 436; nobility, 438; warfare, 440; knighthood, 441; schools, 442; early inhabitants; 449, 450; migrations of the Mexicans, 456; foundation of the city, 462; first king, 466; floating gardens, 469; death of first king, 470; second king, 471; third king, 473; power of kings, 474; murder of third king, 475; fourth king, 476; coronation, 468; 477; Mexico was founded, with the name of Tenochtitlan, in 1327, on a lake in the midst of a valley forty leagues round. Cortes took the city on 13 August 1521. Water brought to, 499; entry of Cortes into, 518; insurrection at, 520; Spaniards retreat from, 521; return of Cortes to, 523
Montezuma I, 487, 488, 493
Montezuma II, 436; character, 500; household of, 503; coronation, 504; Government of, 505; signs and evil omens presaging fall of, 506 to 512; news of the Spaniards, 513; embassy, 514; his terror, 516; strategy, 517; submission to Cortes, 518; and death, 520
Nauincopa, 216
Navatlacas. Invaders of Mexico, 451
Quahuanahuac, 453
Quaxutatlan, 497
Quetzalcoatl, 384, 508, 514
Suchimilcos, 452
Tacuba, 491
Tenoxtitlan (name of Mexico), 478, 480
Tepeaca, 504
Tepanecas, 452, 460, 464, 468, 478, 480, 481, 485
Tepotzotlan, 446
Teuculhuacan, 455
Tezcatlipuca (Mexican God), 339, 377, 378, 379, 517
Tezcuco, 253, 437, 452, 466, 476, 487. A town fifteen miles E.N.E. of Mexico, at the foot of the hills in 19° 31′ 30″ N. Speech of king of, 501, 502
Ticocic, 493

Tiçaapan, 460
Tlatellulco, 496
Tocci, 461. An idol.
Tozcoatl, 377
Tlacael, 436
Tlacaellel, 478, 479, 481, 484, 487; refuses the crown, 491; death, 494
Tlascala, 504, 517, 519, 530
Topilcin, 514
Tula, 459

Vitzilipuztli (Mexican god), 305, 356, 455, 457, 460, 463, 469; festival of, 357, 377, 491, 500
Vitzilovitli, 471
Zacatecas, 210. A province in the north of Mexico on the tropic of Cancer, 210 miles long by 177. Its tableland is 6,500 feet above the sea.

VII.

LIST OF SPANIARDS AND OTHER EUROPEANS MENTIONED BY ACOSTA.

Acosta, Bernardo de, brother of the author, i; in Mexico, ix
────── Christoval de, i; author of a work on the drugs of India, ii, *n*.
────── Joseph de, THE AUTHOR, his birth, i; becomes a Jesuit, ii; sails for Peru, ii, 56; on the isthmus of Panama, ii, 263; observes the antics of monkeys, 285; crosses the line, iii, 90; arrival in Peru, iii; crosses the Andes, his sufferings, 130, 131; cured of snow blindness, 288; goes to Lima, v; at the Council of Lima, vii; his sermon, vii; sailed for Mexico, viii, 127, 391, 400; his return to Spain, ix, 194, 204, 226, 239, 260, 271; had seen the part of the heavens unknown to the ancients, 4; his views respecting the peopling of America, 46; believed that the Old and New World were joined, or approached near, 60; heard about the Amazons from a Jesuit who had been with Ursua, 82; saw the comet of 1577 in Peru, 122; saw camels in Peru, 272; knew a man who lost his toes by frost-bite, 133; conversations with Sarmiento's pilots, 140; saw giants' bones in Mexico, 454; his publication of the first two books of the *Natural History* in Latin, ix, xi; his religious works published at Rome, x; his work, *De Promulgatione Evangelii*, xi; his *Natural History* published in Spanish, xii; editions and translations, xiii; the English translation, xiv; account of, by Antonio, xv, *n.*; his death at Salamanca, x
Aguirre, Lope de, the famous pirate who descended the river Amazon in 1560. Acosta heard of the wonderful voyage from a Jesuit who, when young, was in it, v, 83. He has Diego instead of Lope. (See *Search for El Dorado*, Hakluyt Society's volume for 1861.)
Alcobaça, Diego de, his confessionaries in native languages, v
Almagro, Diego de, allusion to his invasion of Chile, 133
Alonzo, Hernando, pilot in the expedition of Sarmiento, his account of the Straits of Magellan, 143
Alvarado, Pedro de, 521. In command at Mexico. He was the chief lieutenant of Hernan Cortes, and afterwards conquered Guatemala.
Antonio, Dr. Nardo, an Italian physician, alluded to as quoting from the work of Dr. Francisco Hernandez, 261
Arriaga, Pablo Jose de, his work on the extirpation of idolatry, v
Avila, Dr. Francisco de, his work on the folk-lore of Huarochiri, v
Balboa, Blasco Nunez de, discoverer of the South Sea, 135
Bertonio, Ludovico, his Aymara dictionary, v
Betanzos, Fray Domingo de, a Dominican, 531. He was born at Leon, and studied at Salamanca, whence he went to Rome to seek permission from the Pope to become a hermit. Having obtained the desired leave, he went to the Isle of Ponza and lived there as a hermit for five years. He then became a Dominican and, in 1514, he went to Hispaniola. In 1526 he was one of the first twelve Dominicans who

went to Mexico. Thence he removed to Guatemela, and, after labouring for many years, he returned to Spain, and died in the monastery of San Pedro at Valladolid.

Cabrera, Amador de, possessor of a rich quicksilver mine at Huancavelica, which he sold, 216

Cañete, Marquis of, 432, 435

Carbajal, Gutierrez, Bishop of Plasencia. A ship of his passed through the Strait of Magellan, 137.

Castro, Lope Garcia de, 215. Governor of Peru, under the title of President of the Audience. He succeeded the Viceroy, Conde de Nieva, who was assassinated in 1562. In his time the quicksilver mines of Huancavelica were discovered. He colonised the island of Chiloe, founding the town of Castro. In 1567 he despatched the expedition, under his young nephew Alvaro de Mendaña, which discovered the Solomon Islands. In 1567 the Jesuits arrived in Peru. Castro was succeeded by the Viceroy Toledo in 1569.

Cavendish Thomas, his capture of a prize near California, 141 n.

Centeno, Diego, a vein of silver ore at Potosi named after him, 199. A man of good family, native of Ciudad Rodrigo. At the age of twenty he came to Peru with Pedro de Alvarado in 1534. He fought on the side of the Pizarros at the battle of Las Salinas on April 26th, 1538, and under Vaca de Castro at Chupas. He received a rich estate at La Plata (Chuquisaca), in the province of Charcas, where he was Alcalde when the Viceroy Blasco Nunez Vela published the new laws. At first he was opposed to them, but he eventually rose against Francisco de Almendras, whom Gonzalo Pizarro had appointed his lieutenant in Charcas. He seized Almendras, who was a friend and almost a brother to him, and had him beheaded at La Plata. Gonzalo Pizarro sent Carbajal against Centano, who defeated him several times, and he was obliged to hide in a cave near Arequipa. On the arrival of Pedro de la Gasca in Peru Centeno again collected a force, but was defeated by Gonzalo Pizarro in the battle of Huarina. He escaped and joined Gasca at Andahuaylas, being present with him at the battle of Sacsahuana. He had charge of the person of Gonzalo Pizarro until his execution. Centeno died in 1549. He was a short fair man, with a red beard.

Columbus Christopher. A nameless pilot said to have given the secret of the discovery of America to, 54. For a full discussion of this story, see my note in the first volume of my translation of the *Royal Commentaries of Garcilasso de la Vega*, p. 24.

Cortez, Hernando, Marques del Valle, conqueror of Mexico, 304, 353, 458, 498; his arrival on the coast of Mexico, 514; march to Mexico, 517, 518; interview with Montezuma, 519; return to Mexico, 523

Costillas, Geronimo, lost his toes from frost bites in Chile, 133. He was a native of Zamora, of good family. He dissuaded Almagro from executing Hernando Pizarro, and fled from Gonzalo Pizarro to Arequipa and Lima. He was afterwards actively engaged in the campaign against Giron. He had a house at Cuzco. (See *G. de la Vega*, ii, p. 243.)

Drake, Sir Francis, vi; his passage of Magellan Strait, 137; his Portuguese pilots land in New Spain, 140; at Guatulco, viii, n.

Ercilla, Alonzo de, 136; said to have written part of his *Araucana* on plantain leaves, 244. For his life and writings, see Ticknor's *Spanish Literature*, ii, p. 426.

Garces, Henrique, a Portuguese, the discoverer of the quicksilver mine of Huancavelica, 215

Gasca, President, 429

Grimston, Edward, English translator of *Acosta*, account of, xiv

Henriquez, Don Martin, Viceroy of Mexico from 1568 to 1580, and of Peru from 1581 to 1583. He was a younger son of the Marquis of Alcanices. (*See* Hawkins's *Voyages*, p. 75, n.) Acosta conversed with him on the subject of a southern continent, vi, 139, 391, 423; his death, vii.

Hernandez, Dr. Francisco, 261. He was born at Toledo in 1514, and graduated at Salamanca. In 1570, Philip II sent him to Mexico, with the cosmographer Francisco Dominguez, to write the natural history of that Viceroyalty. He returned in 1576, but died before he could publish the results of his labours. He prepared sixteen MS. folio volumes, six describing the plants, animals, and minerals of New Spain, and ten of drawings. *Francisci Hernandez rerum medicarum Novæ Hispaniæ Thesaurus seu plantarum, animalium, mineralium, Mexicanorum Historia*, tom i, 1648; ii, 1651, folio. He also translated Pliny's *Natural History*. (*See* Antonio, *Bib. Script. Hisp.*, i, p. 432

Holguin, Dr. Gonzalo, his Quichua grammar, v

Ladrillero, Captain, his account of a voyage through the Straits of Magellan, 137

Lancero,. a soldier of Peru, cures wrought by, 525

Loaysa, Dr. Don Geronimo de, first Archbishop of Lima, 425; letter from Polo de Ondegardo on the rites of the Peruvians, 356. Loaysa was native of Truxillo in Estremadura, a Dominican, made Bishop of Carthagena in 1537. In 1543 he was translated to Lima, which was made an Archbishopric in 1548. He died at Lima in 1575, and was buried in the hospital of Santa Ana, which he had founded.

Magellan, Fernando, his discovery of the Strait, 136

Mandana, Alvaro de, discovered the Solomon Islands in 1568, 46, 115

Matienza, Judge, iv

Melendez, the Adelantado Pedro, affirmed that there was a passage north of Florida, 140; on whale fishing, 150; or Menendez ? Pedro Menendez was a native of Aviles near Oviedo, of a very ancient Asturian family. He was a daring sea captain. In 1565 Philip II sent him with a fleet to conquer Florida. He returned, and died at Santander in 1574. His nephew Pedro, Marquis of Aviles, went out to Florida with his uncle the Adelantado. He was killed by the Indians. Menendez wrote a report on his examination of the east coast of Florida.

Mendoza, Garcia de, Governor of Chile, sent a ship to explore towards the Strait of Magellan, 137. Garcia Hurtado de Mendoza, son of the Marquis of Cañete, was a young man of twenty-two when he came to govern Chile in 1577, sent by his father the Viceroy of Peru. He made a successful war on the Araucanians, and explored the archipelago of Chiloe. He founded Mendoza on the east side of the Andes, and rebuilt Angol and other towns previously abandoned. In 1561 he was superseded and returned to Spain. He came out as Viceroy of Peru in 1590 until 1599. His life by Christoval Suarez de Figueron was published in 1613. (*See* Hawkins's *Voyages*, xxviii, 255, n, 338, 340. Hakluyt Society's volume for 1878.)

Mogrovejo, Dr. Toribio, Archbishop of Lima, vi; lives of, viii, n.

Monardes, Dr., on whale fishery, 150; on liquid amber, 259; on tobacco, 261

Narvaez, Pamphilo, landing in Mexico, 520

Pizarro, Francisco, conqueror of Peru, treasure seized by, 325, 432, 435, 529

Pizarro, Gonzalo, 429

Polo de Ondegardo, iv, v, 304, 313, 314, 356, 369, 391, 425, 432, 434, 525, the licentiate, was born at Salamanca, and in 1545 he was in Peru, with the fame of a very learned and prudent man. He was a friend of Gonzalo Pizarro, yet Gasca made him corregidor of Potosi. Afterwards he was corregidor of Cuzco, when he discovered several mummies of the Yncas, which were sent to Lima. He was the adviser of the Viceroy Toledo, and died at Potosi in about 1575, very old and rich. His widow married Don Alonzo de Loaysa, a citizen of Potosi, and survived until 1603. His valuable *Relaciones* are addressed to the Viceroys Marques de Cañete and Conde de Nieva, 1561-71. They show him to have been a humane and good man. They are in MS. in the Escurial. Another MS. of Ondegardo is in the Royal Library at Madrid. It is printed in the Hakluyt Society's volume for 1872, p. 151. (See also Prescott's *Conquest of Peru*, i, p. 163.)

Roa, Juan de, an Austin friar, and zealous preacher, 531

Salinas, Juan de, the Adelantado, his entry of the river Amazons, 157

Sanchez, Father Alonzo. On the trade winds, 123 ; Chinese writings, 400

Sarmiento, Pedro de Gamboa, vi, 137, 138. Sarmiento had studied the records and ancient traditions of the Yncas, one of which told how the Ynca Tupac Yupanqui had visited the islands far to the west, called Ahuachumbi and Ninachumbi. He sailed in the fleet of Alvoro de Mendaña in 1567, with the object of reaching these islands. He is believed to have written a *Historia de los Yncas*. In 1579 he was sent with a fleet from Lima to explore the strait of Magellan. His journal was published at Madrid in 1768. *Viaje al estrecho de Magallanes por el Capitan Pedro Sarmiento de Gamboa en los anos 1579 y 1580*. There is an account of Sarmiento and his surveys in Burney's *Voyages*, ii, pp. 3 to 57

Tobar, Juan de, ix, 391

Toledo, Francisco de, second son of the Count of Oropesa. Viceroy of Peru from 1569 to 1581, iii, iv, 137, 151, 204, 216, 231, 256

Torres, Rodrigo de. A miller who introduced the use of *ychu* grass for fuel, in mining, 218

Treço, Tacomo de, of Milan. A worker in brass at Madrid. The way his workmen preserved themselves from the injurious effects of the fumes of quicksilver, 212

Ursua, Pedro de, commander of the expedition down the Amazon in 1560. (*See* Aguirre) 157, 171

Vaca, Cabeza de, 525. In 1527 he went as treasurer in the expedition of Pamphilo de Narvaez to Florida. Narvaez was lost in a storm near the mouth of the Mississippi, and Cabeza de Vaca took command. He and his followers were reduced to the necessity of cannibalism, and were afterwards made slaves by the Indians. He escaped, and, after passing through a variety of incredible hardships, reached Mexico. He retired to Spain in 1537. He was afterwards Governor of Paraguay.

Valdes, Diego Flores de. The officer sent, with Sarmiento, to fortify Megellan's straits (*see* Sarmiento), 139

Valencia, Fray Martin de, 531. A zealous Franciscan preacher. He was a native of Valencia. In 1523 he was appointed to take out twelve Franciscans to Mexico, as their provincial. Here he worked zealously for the conversion of the Indians. He died on a journey from Mexico to Tehuantepec, on August 31st, 1534. He wrote interesting letters to Charles V and to the Pope Adrian VI, as well as to Friar Matthew Weiser, the General of his Order, describing the spiritual conquest of Mexico. He was also the author of some historical documents.

Valera, Blas, the Jesuit, v.

Valle, Marques del. (*See* Cortes.)

Velasco, Pedro Fernandez de, introduced the refining of silver with mercury in 1571, 217

Villaroel, the Spaniard whose servant discovered the mines of Potosi, 203

CPSIA information can be obtained
at www.ICGtesting.com
Printed in the USA
FSHW012116161219
65171FS